# SOIL SCIENCE SIMPLIFIED
## FOURTH EDITION

# SOIL SCIENCE SIMPLIFIED
## FOURTH EDITION

**HELMUT KOHNKE**
*late of Purdue University*

**D. P. FRANZMEIER**
*Purdue University*

WAVELAND
PRESS, INC.
Long Grove, Illinois

For information about this book, contact:
Waveland Press, Inc.
4180 IL Route 83, Suite 101
Long Grove, IL 60047-9580
(847) 634-0081
info@waveland.com
www.waveland.com

**Cover:** The cover sketch represents an older, nostalgic view of soils. Now farmers do not use a one-horse plow, and they do not leave sloping soils bare. Illustration by Lou Jones, Agronomy Department, Purdue University.

10-digit ISBN 0-88133-813-3
13-digit ISBN 978-0-88133-813-3

Printed in the United States of America

19   18   17   16

# CONTENTS

# PREFACE

This book is a distillation of soil science, written for those who want to get acquainted with the basic concepts of soils but do not have the time nor the inclination for extensive study. Previous editions--published in 1953, 1962 and 1966--have found wide acceptance in colleges and high schools, and by farmers, agronomists and many others.

Dr. Helmut Kohnke wrote the previous editions. He was born in Russia of German parents, graduated from the University of Berlin with a doctorate degree, and earned advanced degrees from the University of Alberta, Edmonton, and Ohio State University. He spent most of his professional career at Purdue University, where he taught and conducted research in soil physics, soil conservation and water management. He also worked for the United States Department of Agriculture Soil Conservation Service and served as a visiting scientist in Germany, Brazil, Argentina, Colombia, Iran and Bulgaria. His hobby was preserving natural areas through the Nature Conservancy and the Wildcat Park Foundation of Indiana. He died in 1991 at the age of eighty-nine. Helmut's thorough scientific training and broad geographic experience, his love of nature, and his attitude of stewardship of natural resources are reflected in the book.

The discussion of the scientific principles of soils in the previous editions has withstood the test of time. This edition builds upon the foundation of the previous ones. The first six chapters and chapter 11 are essentially from the third edition. Chapters 7 through 10 are mainly new. They reflect the modern system of soil classification and explain how people can use published soil surveys to help them understand their soil resources and how to manage them. Chapter 12, "Soil and the Environment," explains how soils relate to their natural environment, how people have harmed them, and why everyone should be concerned with how we treat this essential natural resource. This chapter will be of interest to those concerned about natural resources, environmental science and the relation of soil to society, including the significance of the rapidly increasing population.

# THE NATURE AND FUNCTION OF SOIL

What is soil? A definition of soil may be given from several viewpoints, depending on the function of the soil in which we are interested. The geologist may consider soil to be the decomposed surface part of the rocks. The engineer may stress the physical characteristics of soil in defining it, for instance its compressibility, its bearing strength, and its permeability to water. To the pedologist soil is a natural body, occurring in various layers, composed of unconsolidated rock fragments and organic matter.

The agronomist defines soil as the unconsolidated cover of the earth, made up of mineral and organic components, water and air and capable of supporting plant growth. The latter definition seems to be the most appropriate one for a farmer or a conservationist, since it includes the most important function of the soil: to grow plants.

The growth of most plants is impossible without soil (fig. 1-1). Soil teams up with sun and rain to provide our food and clothing. Besides fish and water cultures, there is no other source of food than the crops grown on soil, and it is obvious that fish and water cultures can provide only a small fraction of the food that people need. Our survival depends on the conservation of the body and the fertility of the soil. We gauge, therefore, the value of the soil by its capacity to produce crops.

A green plant has the ability to combine carbon dioxide and water from the ground into sugar, other carbohydrates and fat by the process called photosynthesis. Light furnishes the energy necessary for this reaction. Nitrogen, sulfur and phosphorus are required for the synthesis of proteins along with carbon, oxygen and hydrogen. Several other elements are needed for essential plant functions.

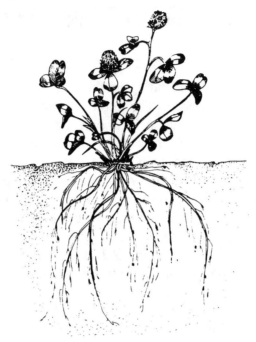

Fig. 1-1. Plant growth is the main reason for our interest in the soil.

A green plant is made up of about 80% water and 20% dry matter. The largest part of the dry matter is composed of the elements hydrogen, oxygen, carbon and nitrogen which occur in air and water, while the rest consists of a great variety of elements that originate in the soil. In spite of the small amounts of these components, they are absolutely essential and the complete absence of only one of them makes plant growth impossible.

As a medium for plant growth, soil performs four functions:

It serves to anchor the roots.

It supplies water to the plant.

It provides air for the plant roots.

It furnishes the minerals for plant nutrition.

How can soil perform this function of storehouse for water, air and plant nutrients and be permeable enough for the tender root to penetrate into it and yet so powerful to protect large trees from being blown over by the wind? Acquaintance with soil formation, the physical and chemical properties of the soil, and with the functions of the microbes of the soil will clarify these mysteries.

Soil consists of solid particles, water, and air and, in addition, contains a teeming population of minute plants and animals. The solids are both mineral and organic. The mineral particles are classified according to their sizes into gravel, sand, silt and clay. The organic matter consists of fresh plant and animal residues, which are readily decomposed, and of more stable humus. These soil particles do not lie disorganized side by side but are usually associated into smaller or larger groups. These aggregates may be small crumbs or large clods.

Of the solids, clay and organic matter are of major importance in the nutrition of plants, since they are chemically active. Gravel, sand and silt are largely inert and contribute little to plant nutrition. Over many years, however, some of these particles weather to release nutrient elements that are held by clay and organic matter.

The pore space between the solids is taken up by water and air. The water might more appropriately be called a solution because it contains small quantities of numerous minerals. These serve as nutrients for the plants.

Air takes up that part of the pore space not occupied by water. As the water content increases, the air content decreases. The plant roots require oxygen for their normal functions just as the above-ground plant parts do and as animals breathe. In respiration, plant roots use oxygen and give off carbon dioxide. For this reason, soil air usually contains less oxygen and more carbon dioxide than atmospheric air does. A continuous replenishment is necessary to keep the oxygen content sufficiently high. Large pores and an intermediate moisture content are helpful for this.

Millions of microbes live in each ounce of fertile soil. Without them soils would be inactive and soon lose their capacity to support plants. Microbes help to bring plant nutrients into available form and they make soil crumbs stable and resistant to erosion. Creating a hospitable environment for microbes in the soil is an important task of the farmer.

Yield and composition of crops depend to a large extent on the properties of the soil. Humans, who eat these crops and the meat from the animals raised on these crops, are truly a product of the soil and reflect in their bodies -- and minds -- the wealth or the poverty of this land.

# 2

# PHYSICAL PROPERTIES OF SOILS

## Soil Components

Soils consist of solid, liquid, gaseous and biotic components. The solid components are mineral and organic. The mineral soil particles are classified according to size as shown in table 2-1.

Table 2-1. Size limits and description of soil fractions.

| Soil fraction | Diameter | Description |
|---|---|---|
| Gravel | Larger than 2 mm | Coarse |
| Sand | 0.05 - 2 mm | Gritty |
| Silt | 0.002 - 0.05 mm | Floury |
| Clay | Smaller than 0.002 mm | Sticky when wet |

Sand and silt are merely broken down rock fragments (fig. 2-1); they consist of quartz, feldspar, mica or other minerals. Chemically they are essentially inert compared with clay and organic matter, which are responsible for most of the chemical reactions of the soil. Sand and silt play an important role by providing a skeleton for the soil.

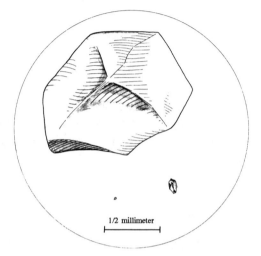

1/2 millimeter

Fig. 2-1. The relative sizes of coarse sand, fine sand and silt. The small speck near the center is the silt. An individual clay particle remains invisible even under this magnification.

Clay particles are plastic and sticky when wet. They are highly adsorptive of water, gas and dissolved substances. Clays are minute, plate-shaped, aluminosilicate crystals, consisting of silicon, aluminum, iron, magnesium, oxygen and hydrogen. They may also contain potassium, calcium and other elements. The biggest clay particle is less than one ten-thousandth of an inch in diameter (0.002 mm). There are several types of clay. Two of the most important ones are kaolinite (fig. 2-2) and smectite (fig. 2-3). The latter is still smaller than kaolinite. Smectite clays have the ability to swell on wetting and to shrink when they are dry. They enter into physical and chemical reactions to a much larger extent than kaolinitic clays.

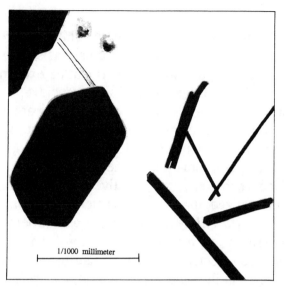

Fig. 2-2. Silhouette view of kaolinitic clay as seen with the electron microscope.

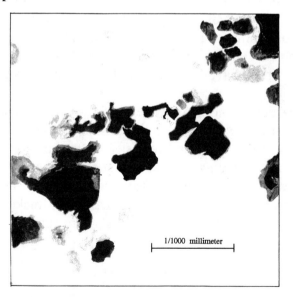

Fig. 2-3. Silhouette view of smectite clay as seen with the electron microscope.

Clay is a negatively charged colloid. This negative charge is the reason that positively charged cations surround each clay particle. The individual cations can be exchanged for each other. The hydrogen ions of slightly acidic percolating water or of the acids given off by plant roots are absorbed by the clay, while calcium, magnesium, potassium and sodium are released in equivalent quantities and become available for plant nutrition. If the cations can get close to the surface of the clay, the negative charge on the clay is largely neutralized and the clay particles will cling together. They are *flocculated*. This is the case where calcium and magnesium are the dominant cations. These ions are small and are effective in holding clay particles together. Sodium ions, on the other hand, are large because they are covered with a shell of water and cannot get close enough to the surface of the clay to effectively neutralize its negative charge. Such clay particles repel each other, and soils containing sodium clay are *dispersed*, that is, they have no tendency to form aggregates. Soils with calcium clay generally have a more desirable structure than sodium soils.

Silt particles hold much water in the soil and most of this water is available for use by plants. Over time, some silt particles break down and release ions to the soil solution, so they also serve as a storehouse for plant nutrients. Sand particles tend to keep the soil loose and, in this way, counteract the tendency of clay particles to make the soil tight and impermeable to water and plant roots.

## Soil Texture

The relative proportion of the various grain sizes in a soil is called *texture*. To describe soil texture, names such as loamy sand, silt loam, clay loam and silty clay are used, as shown in figure 2-4. The best soils are generally those which contain 10 to 20% clay, with silt and sand in approximately equal amounts, and a fair amount of organic matter.

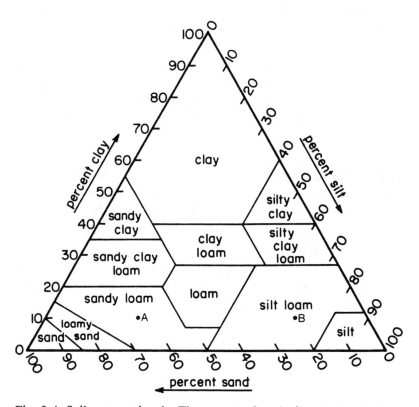

Fig. 2-4. Soil texture triangle. The content of sand, silt and clay for the twelve main soil texture classes can be found on this triangle. For example, the soil represented by point A is in the sandy loam texture class with 65% sand, 25% silt and 10% clay. The one at B is a silt loam with 10% sand, 70% silt and 20% clay. Note that soils with relatively small clay contents (< 40%) are in the clay texture class because the properties of clay readily predominate over the coarser fractions.

## Surface Area

In comparing clay with sand and silt, it is important to be aware of the relative amount of surface area of these particle-size groups, because it is on the surface that many chemical and physical processes take place. The surface area of one gram (1/8 teaspoon) of spherical particles of different diameters is shown in table 2-2. One gram of sand-size particles has a surface area the equivalent of three lines on this page; one gram of silt has a surface area about the same as four pages of this book; and one gram of clay-size spheres has about the same surface area as the wall of a large room in a house. Also, soil clays are plate-shaped instead of spherical, and they may be smaller than 0.0002 mm and have internal surface area; therefore, the surface area of one gram of soil clay can be as large as the surface area of all the walls, floors, and ceilings of a house. Organic matter also has a very large surface area. It is apparent that the greater the surface a substance exposes, the greater will be its ability to enter into chemical and physical reactions.

Table 2-2.    Number of spherical particles and total surface area in one gram of material with different particle size.

| Kind of particle | Diameter of particle | Number of particles in 1 gram | Surface area of one gram |
|---|---|---|---|
| Sand | 2 mm | 90 | 11 cm$^2$ |
| Silt | 0.02 mm | 90,000,000 $(9 \times 10^7)$ | 1130 cm$^2$ |
| Clay | 0.0002 mm | $9 \times 10^{13}$ | 113,000 cm$^2$ |

## Soil Structure

One of the most important physical properties of a soil is the arrangement of its individual particles in relation to each other, or its *structure*. There is an infinite number of possibilities in which the particles can be arranged.

Soil structure is the arrangement of particles into small groups, or aggregates. These aggregates may be bound together with other aggregates into larger masses called peds (fig. 2-5a). The peds come in different shapes that roughly resemble spheres, blocks, columns and plates. They may have rounded or sharp edges and corners. The amount of pore space within an aggregate depends mainly on the soil texture, and the amount of pore space between the aggregates depends on their arrangement with respect to each other, much as the size of the rooms of a house depends on the arrangement of the walls. If the individual particles are arranged in small aggregates with rounded edges, we speak of *granular* structure. This is very desirable for plant growth, because it provides both large and small pores.

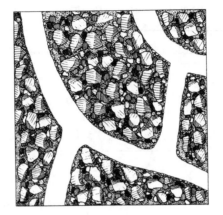

(a) Granulated soil -- Granular structure, aggregated particles.

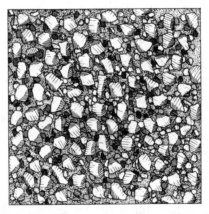

(b) Puddled soil -- Massive structure, dispersed particles.

Fig. 2-5. Schematic cross sections (a) through a porous well-aggregated soil, and (b) through a tight puddled soil. In a puddled soil there are only small pores, but in an aggregated soil there are also large pores that drain readily. Cross sections enlarged 30-fold.

Some soils lack structure. In very sandy soils the individual grains act independently of each other. They lie in a random distribution, with smaller grains filling the openings between the larger grains so that a minimum of pore space exists. No binding substances hold the particles together, so the soil has no peds. If some soils are tilled too much or driven over when they are wet, their natural structure may be destroyed and they become puddled (fig. 2-5b). Their structure is called *massive*.

Aggregation of the individual soil particles occurs through various natural agencies, such as root growth and decay, microbial activity or swelling and shrinking of clay, which may be caused by wetting and drying or freezing and thawing. Also, iron and aluminum oxides serve to cement particles together.

The growth and decay of grasses and legumes particularly stimulate aggregation. Soil can only aggregate if clay or organic matter -- preferably both -- are present. Aggregates formed without organic matter break down when submerged in water for some time; only aggregates that are held together by organic matter in some form are water-stable. Soils that are in optimum condition of structure and are made up of water-stable aggregates are said to be in good *tilth*.

## Pore Space

The pore space determines the amounts of air and water contained in the soil and it also is an important factor in root penetration, so it deserves close attention.

Large pores (larger than 0.06 mm diameter) are readily drained of water and filled by air after a heavy rain. They are valuable as an aeration system. Small pores (smaller than 0.06 mm diameter) hold water against gravity and, in fact, pull water up from a water table by capillary action. Therefore, they are necessary for the water supply of plants. The structure is ideal where large and small pores occur in a proportion that corresponds to the water and air needs of the crop plants in a given climate and under the cultural conditions employed. Under ordinary farming conditions in the humid temperate zone the ratio between the volume of the large and small pores should be about 1:1.

## Soil Temperature

The soil temperature is as important to plant growth as air temperature. The temperature of the surface soil fluctuates greatly both during a twenty-four hour period and with the seasons. The farther down in the ground the temperature is measured, the smaller are the fluctuations. At two to three feet, daily changes no longer occur, and at about twenty-feet depth the soil temperature remains the same during winter and summer, at about the average annual temperature. Where the soil is covered by a dense growth of plants or a thick layer of mulch (plant residues), temperature variations are much less severe and do not penetrate as deeply. Soil temperature has a direct effect on plant growth and also influences microbial activity (e.g., nitrification). Freezing and thawing of the soil water also affects soil structure. Slow and occasional freezing and thawing -- as under an insulating layer of mulch -- is beneficial for soil structure. Under rapid freezing and thawing, however, soil aggregates are generally broken down too much for good tilth. Besides, frost action alone does not create water-stable aggregates.

Moisture movement, especially in the vapor phase, is brought about by temperature differences in the soil or between the soil and the air. When we speak of a cold soil we mean a soil that retains much water in the spring and warms up slowly because of the high specific heat of water as compared to that of air, which would otherwise fill the pores.

## Soil Air

Air in the soil is needed for the respiration of plant roots and microbes. It is also necessary for making nutrients available to plants. Without air in the soil, few plants can survive. Soil air usually contains more carbon dioxide and less oxygen than atmospheric air because oxygen is constantly used up and is only replenished by movement of air or by diffusion. Air fills all pore spaces in the soil not filled by water; the wetter the soil the less air it contains. As the larger pores are most readily freed of water and filled by air, the amount of large-pore space is called *aeration capacity*. Sandy soils and those that are well aggregated have large aeration capacities.

## Soil Color

The color of the soil tells us much about some of its other properties. The color of a surface soil horizon depends mainly on its organic matter content -- the darker the soil, the more organic matter it contains. This organic matter imparts favorable properties to the soil, such as better aggregation and a higher water-holding capacity. Also, dark soils absorb more radiation during the day than light-colored soils and radiate more heat during the night.

In subsoil horizons, soil color indicates the wetness and aeration condition of the soil. In general, reddish and brownish subsoils indicate good aeration and little water logging. Grayish and olive colors indicate much water logging and chemical reduction of iron. A mottled subsoil, one with a splotchy pattern of brownish and grayish colors, is indicative of a fluctuating ground water table. These properties are important for plant growth because plants need oxygen. A soil that is severely reduced and gray will have long periods when oxygen is deficient, so the plants will suffer.

# 3

# SOIL AND WATER

## Hydrology

Hydrology is the study of the water cycle in nature. The ever recurring conversion of ocean water to atmospheric water, to precipitation, to ground water, to runoff, and to ocean water with its many ramifications is called the water cycle or *hydrologic cycle* (fig. 3-1).

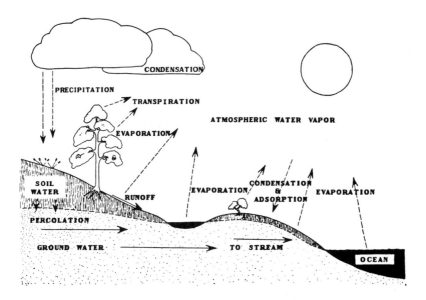

Fig. 3-1. The water cycle.

Plants are on liquid diets; they absorb most of their nutrients from the soil solution. That makes plants very heavy drinkers. They use from 200 to 1000 pounds of water for the formation of one pound of dry matter. This ratio is called the *water requirement*. The water requirement varies with the site and climate. In fertile soil and in a moist climate where growing conditions are good, the amount of water needed to produce one pound of dry matter is much less than in poor soil with inadequate moisture supply. The amount of water that passes through the corn plants of one acre to make a 125-bushel crop corresponds to approximately 16 inches of rain.

This great need for water by plants makes it mandatory that sufficient water be stored by the soil. Infiltration must be encouraged and surface runoff must be kept to a minimum. This is also necessary in order to avoid excessive erosion.

## Energy of Soil Moisture Retention

Water is held on the surface of each soil particle and between the particles in the pore space. Since clay has a much higher surface area (area per unit weight) than silt or sand, it holds much more water. To put it another way, a certain amount of water is held much more tightly by clayey soils than by sandy soils. This is so because water is held tightest near the surface of a soil particle and less tightly farther away. Comparing sand and clay with the same moisture content, the layer of water covering an individual sand particle is much thicker than the layer of water covering a single clay particle. Therefore, the amount of water that a soil can release to plant roots depends on the energy with which the water is held by the soil, rather than on the total amount of water present.

Because the surfaces of soil particles hold water tightly, plant roots must expend energy to remove water from a soil. Plants can easily utilize water held in medium-sized pores, where it is relatively far from soil particle surfaces, but it is more difficult for them to remove water from near the surface of a clay particle. The process of exerting energy to remove water from a soil is simulated in the laboratory by applying suction to the moist soil. In the field and in the laboratory, the energy required to remove water from a soil is called the *soil matric potential* and is measured in kilopascal (kPa) units (table 3-1). When all the pores of a soil are filled with water, the soil is said to be *saturated* and the matric potential is zero. A few days after a soaking rain, water has drained from the larger pores and is held only in the medium-sized and small pores. This soil is at *field capacity (FC)*, and the water is held at around 10 to 33 kPa. When a plant has removed all the water it can from a soil and the plant wilts, water is held only in the smallest pores and as thin films on soil particles, and the soil is at its *permanent wilting point (PWP)*, which corresponds to about 1,500 kPa. Some water is held more tightly than 1,500 kPa, and this water is not available to plants. For comparison, 1,500 kPa equals 217 pounds per square inch pressure.

Table 3-1.   Soil-water relations. Hygroscopic water is absorbed from humid air and is held very tightly; capillary water is held in small pores with moderate energy; and gravitational water is loosely held and drains by gravity.

| Soil water reference points | | Soil water ranges | |
| --- | --- | --- | --- |
| Description | Soil matric potential | Soil appearance | Plant-water relations |
| Oven dry | ---- | | |
| | | Dry | Too dry for plants |
| Permanent wilting point | 1,500 kPa | | |
| | | Moist | Water available to plants |
| Field capacity | 33 kPa | | |
| | | Wet | Not enough air for plants |
| Saturation | 0 kPa | | |

Water held between FC and PWP can be used by plants and is called the *available water capacity (AWC)* of the soil. In general, sand holds little water at PWP, and a bit more at FC, and has a low AWC. Silt holds much water at FC, little at PWP, and has a large AWC. Clay holds much water at both FC and PWP and has a moderately low AWC. At water contents higher than FC, the soil appears wet; between FC and PWP, it appears moist; and below PWP, it appears dry (table 3-1).

## Significance of Soil Moisture Conditions

Drainage lines placed at a depth of three to four feet remove water from large pores, but do not remove enough water to bring a wet soil down to field capacity. After the drain line has removed all the water it can, only the larger pores (larger than 0.06 mm in diameter) of the surface soil are freed of water and filled with air.

A moist soil contains water and air in proportions that are conducive to the growth of crop plants and microbes. In this range the soil is *friable* (easily crumbled with the fingers) and can readily be worked by machinery without being puddled or powdered into a poor structure. Soil water in the moist range, between FC and PWP, moves from wetter to drier locations through very fine, hairlike pores essentially independent of gravity. These pores, called capillary pores (*capilla*, Latin for hair), are illustrated in figure 3-2.

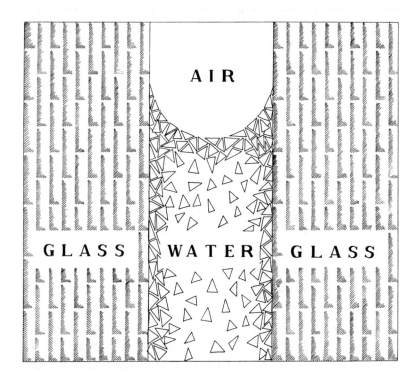

Fig. 3-2. Surface tension lifts water in a glass tube or in the soil. The water molecules (represented by the triangles) are attracted more by the glass walls than by the other water molecules. Capillary moisture movement goes from the wetter to the drier part of the soil, regardless of the direction (up, down or sideways).

When the soil is in the dry range it still contains some water, but this is held with such a force that plants can no longer make use of it. Fine-textured soils (for example, clay loams and silty clays) are quite hard in this condition and resist the penetration by tillage implements, especially if the soils are deficient in organic matter. Dry soils can absorb moisture from the atmosphere if the vapor pressure gradient is favorable (if the atmosphere is relatively more moist than the soil).

## Measuring Soil Moisture

The moisture status of a soil can be determined according to two principles. Where it is desired to determine the amount of water in the soil, a sample is weighed, dried to constant weight in an oven at 105°C (just above the boiling point of water) and weighed again. The difference in weight is ascribed to the moisture driven off and this is then referred to either the weight or the volume of the original sample. The percent of water by weight is calculated by the following relationship:

$$\% \text{ water} = \frac{\text{wt. (moist) - wt. (oven dry)}}{\text{wt. (oven dry)}} \times 100$$

Water is not always determined by oven-drying, but this method is used as the standard to calibrate other techniques. The bulk density of soil is the dry soil weight divided by the volume of soil, including particles and pore space. It is often expressed in units of $g/cm^3$. The percent of water by volume is equal to the percent water by weight multiplied by the bulk density of the soil.

If it is desired to determine the energy with which the water is held by the soil, various methods are employed that use tension, pressure, electric conductance, vapor pressure and the freezing point of the soil solution according to the moisture range of the soil in question. It is possible to determine the relationships between the amount of soil moisture and the energy of its retention for every soil.

## Soil Moisture Management

It is the aim of soil water management to provide sufficient water required by the crop plants for optimum growth without endangering the health of the plants by a lack of soil air. Moisture needs during the main growth period are usually in excess of the amount of rain falling at that time; therefore, the soil must act as a storehouse and carry water over from the dormant season or dry period. To make this possible two requirements must be met: (1) The precipitation that falls must enter the soil in a process called *infiltration*, and (2) the soil must have a large water-holding capacity to be able to retain much water. Both these requirements call for a well-aggregated, open soil that does not lose its structure too readily. Soil in this condition also permits drainage of excessive water and plentiful aeration of the pores. Soil in "good tilth," as just described, has the additional advantage of having only limited surface runoff and therefore little erosion.

There are other methods of soil moisture management. Excessive moisture can be removed by subsurface drain lines, ditches or surface furrows. Additional water may be supplied by irrigation. Also, surface water may be slowed down or arrested in its course to the river by contour cultivation, terracing or creating small depressions in the soil surface. These practices allow more water to infiltrate into the soil. Crop residue may be left on the ground for the same purpose, as well as to decrease evaporation from the soil surface.

There are, of course, limits to which any of these methods may be successful. Where it is impossible to reach the ideal moisture conditions for one crop, it is wise to use another more adapted one. Winter grain grows during the part of the year when soil moisture is generally high and is better adapted to dry sites than corn or soybeans. If it is impossible to drain a soil adequately, it is better to be satisfied with a moderate yield of crops that can tolerate wet conditions (such as alsike, red top and timothy hay) than to risk losing the entire stand (as could occur with alfalfa).

# 4

# CHEMICAL PROPERTIES OF SOILS

## Chemical Composition of Soils

We know that plants require certain chemicals from the soil, so it is important to be acquainted with the chemical composition of a soil in order to be able to gauge its crop-producing capacity and to determine what adjustments may have to be made. An overall (total) analysis of a surface soil is given in table 4-1.

Five elements account for 95% of the weight of a soil. Oxygen alone represents half of its weight, and because of its large size and relatively small weight, oxygen also makes up more than 95% of the entire volume of the soil. The other elements take up so little space that they fit between the oxygen atoms.

## Availability of Chemicals to Plants

By far the greatest part of the soil is made up of chemicals that are needed by plants in very small quantities or not at all. Potassium, an important plant nutrient, occurs in rather large amounts; however, we notice that plants respond to potassium fertilization on many soils. The reason is simply that the largest portion of soil potassium is contained in rock fragments or minerals that are practically insoluble in soil solutions and, therefore, are of no

Table 4-1.    Average composition of a mineral soil of the northeastern
U.S.

| Element | Symbol | Percent by weight in soil |
|---------|--------|---------------------------|
| Oxygen | O | 50.0 |
| Silicon | Si | 36.5 |
| Aluminum | Al | 5.3 |
| Iron | Fe | 2.0 |
| Potassium | K | 1.7 |
| Carbon | C | 1.0 |
| Sodium | Na | 0.8 |
| Calcium | Ca | 0.7 |
| Magnesium | Mg | 0.5 |
| Titanium | Ti | 0.4 |
| Hydrogen | H | 0.2 |
| Manganese | Mn | 0.1 |
| Nitrogen | N | 0.1 |
| Phosphorus | P | 0.05 |
| Sulfur | S | 0.02 |
| Other elements | | Trace amounts |

immediate value to plants. This situation is similar to that of many other plant nutrients and, therefore, rather than determine how much of a chemical element exists in a soil, it appears more reasonable to determine how much of a chemical element is actually available for consumption by plants. Table 4-2 gives an idea of how small a fraction of the soil components is in a form available to plants. The data are merely broad averages and vary greatly from soil to soil.

Table 4-2. Soil components available to plants.

| Element | Symbol | Amount available | | Ratio of total to available form of element |
| | | Percent of soil | Pounds per acre of plow layer | |
| --- | --- | --- | --- | --- |
| Nitrogen | N | 0.006 | 120 | 17 |
| Potassium | K | 0.0083 | 166 | 2000 |
| Calcium | Ca | 0.10 | 2000 | 7 |
| Phosphorus | P | 0.0017 | 35 | 25 |
| Magnesium | Mg | 0.018 | 360 | 27 |
| Sulfur | S | 0.0012 | 24 | 17 |
| Iron | Fe | 0.0001 | 2 | 20,000 |
| Manganese | Mn | 0.002 | 40 | 60 |

Numerous methods have been developed to estimate the amount of available plant nutrients in the soil, especially for those elements that frequently occur in quantities insufficient for full crop yields, i.e., phosphorus and potassium. It is naturally difficult to imitate the action of roots over a period of several months by a chemical treatment that must of necessity be of short duration. Extractions with acids of various concentrations and other chemical solutions have been used for the determination of available plant nutrients with a fair degree of success.

## Soil Reaction

Soil reaction is expressed as pH (pH $=-\log[H^+]$), where pH less than 7 represents an acid soil, pH equal to 7 represents a neutral soil, and pH greater than 7 represents alkaline conditions. It is of great importance to the growth of crop plants that the soil reaction (the condition of acidity or alkalinity) be maintained at the proper level. Most crops thrive best in a very slightly acid soil (pH 6.5-6.8), while some plants such as alfalfa and sweet clover prefer neutral soils (pH 7.0), and other plants thrive in definitely acid media (pH 4.0-5.5). Soil reaction (pH) can be determined with color indicators or more accurately by electrometric methods in the laboratory. Soil acidity (low pH) is caused by an excess of hydrogen ions attached to the clay particles. This condition can be remedied by the addition of lime (calcium carbonate) as indicated by the equation below, which shows the reaction of $CaCO_3$ with a clay particle that has two negative charges.

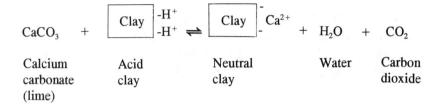

$$CaCO_3 \quad + \quad \boxed{Clay}\begin{array}{l}-H^+\\-H^+\end{array} \rightleftharpoons \boxed{Clay}\begin{array}{l}Ca^{2+}\\-\end{array}^- \quad + \quad H_2O \quad + \quad CO_2$$

| Calcium carbonate (lime) | Acid clay | Neutral clay | Water | Carbon dioxide |

This reaction demonstrates that if lime is brought together with acid clay, neutral calcium clay, water and carbon dioxide are formed. In soil testing laboratories, soil pH is first determined and if it is less than some critical value, such as 6.5, a further test for lime requirement is made to determine the amount of lime that must be added to the soil.

## Cations Associated with Clay and Organic Matter

Because clay and organic matter are the chemically active ingredients of soil, specific attention is given to their composition and reaction. There are several types of clay that act quite differently chemically as well as physically. The cations (ions with positive charge) associated with the clay and organic matter, *exchangeable cations*, are of considerable importance. A prevalence of calcium is best for soil fertility and soil structure. Too much hydrogen makes the soil acid  and releases aluminum compounds to the soil that are toxic to many plants. Relatively small amounts of sodium make the soil too alkaline and cause it to disperse and become impervious and sticky.

### Anions

Phosphate, sulfate, nitrate, chloride and bicarbonate are the important anions (ions with negative charge) of the soil. The first four are essential plant nutrients, while bicarbonate helps to disintegrate soil minerals and to bring plant nutrients into available form. Phosphate is held tightly by soil colloids and cannot be washed out of any but the sandiest soils. Sulfate, nitrate, chloride and bicarbonate are not adsorbed by clay and move freely with the soil water.

## Soluble Salts

There is a continuous formation of soluble cations and anions in the soil, especially during the warm season, as a consequence of weathering of soil and rock particles and the decomposition of organic matter. Where these ions are not removed by leaching (washing them out with water), e.g., in dry climates and in greenhouses, an accumulation of salts occurs that may inhibit plant growth. It is simple to determine the presence of excess amounts of salts in the soil by evaporating and weighing a water extract or by determining the electrical conductivity of the moist soil or of a soil extract. It is not always as simple to remedy the situation because the water required to wash the salts out may not be available. A soil containing over 0.2% soluble salts is generally too saline for optimum plant growth.

# 5

# PLANT GROWTH AND PLANT NUTRITION

## Composition and Life Functions of Plants

Plants are made up of a wide array of chemical compounds. The average composition of a green plant is:

Water . . . . . . . . . . . . . . . 80%
Carbohydrates and fat . . . . . 14%
Protein . . . . . . . . . . . . . 4%
Minerals . . . . . . . . . . . . 2%

Five functions are essential to plant life:

1. *Absorption of water and nutrients* by the roots and to a limited extent through the leaves.
2. *Transpiration* of water from the plants (mostly the leaves) into the atmosphere. Great amounts of water are required for the plant to take up minerals from the soil and to sustain its other life functions.
3. *Photosynthesis*, the creation of plant material through the chemical combination of carbon dioxide of the atmosphere and water of the soil:

$$6\,CO_2 \;+\; 6\,H_2O \;\xrightarrow{\text{energy absorbed}}\; C_6H_{12}O_6 \;+\; 6\,O_2$$

Carbon dioxide    Water    Sugar    Oxygen

This synthesis is possible only in the presence of light and with green chlorophyll as the activating agent.

4. *Synthesis of complex organic compounds.* Carbohydrates, fats, protein, lignin and many other compounds are formed from simple sugars, nitrogen compounds and minerals (salts). The energy required for this synthesis is produced in the respiration process.

5. *Respiration.* Like all living things, plants breathe. Both their tops and their roots inhale oxygen and exhale carbon dioxide. Chemically, respiration is the reverse process of photosynthesis.

energy released

$$C_6H_{12}O_6 \;+\; 6\,O_2 \;\rightarrow\; 6\,CO_2 \;+\; 6\,H_2O$$

| Sugar | Oxygen | Carbon dioxide | Water |

Except for carbon dioxide for photosynthesis and oxygen for respiration by the above-ground parts, the largest portion of the other plant nutrients enters the plants through the roots and must come from the soil. This includes water, oxygen for respiration of the roots, and minerals. Table 5-1 shows the relative amounts of the individual elements in a corn plant.

Table 5-1.   Elementary composition of a corn plant (entire plant).

| Element | | Percent of total dry weight | Element | | Percent of total dry weight |
|---|---|---|---|---|---|
| Oxygen | O | 44.5 | Phosphorus | P | 0.20 |
| Carbon | C | 43.6 | Sulfur | S | 0.14 |
| Hydrogen | H | 6.3 | Aluminum | Al | 0.10 |
| Nitrogen | N | 1.25 | Iron | Fe | 0.04 |
| Potassium | K | 1.20 | Manganese | Mn | 0.003 |
| Silicon | Si | 1.20 | Zinc | Zn | 0.002 |
| Chlorine | Cl | 0.40 | Copper | Cu | 0.001 |
| Magnesium | Mg | 0.25 | Boron | B | 0.0007 |
| Calcium | Ca | 0.23 | Molybdenum | Mo | 0.0006 |

## Plant Food Elements

Not all the elements that occur in plants are necessary for their life processes. While fair amounts of silicon and aluminum are found in most plants, it has never been proved that their presence is essential. Soils are largely made up of silicon and aluminum, and all soil solutions contain these elements. While future research may show that minute quantities of other elements are needed for plant growth, so far seventeen elements have been found to be essential components of plants (table 5-2).

Table 5-2.   Elements essential for plant growth.

| Used in large amounts | | Used in small amounts from soil solids |
|---|---|---|
| Mostly from air and water | From soil solids | |
| Oxygen (O) | Nitrogen (N) | Chlorine (Cl) |
| Carbon (C) | Phosphorus (P) | Iron (Fe) |
| Hydrogen (H) | Potassium (K) | Manganese (Mn) |
| | Calcium (Ca) | Boron (B) |
| | Magnesium (Mg) | Copper (Cu) |
| | Sulfur (S) | Zinc (Zn) |
| | | Molybdenum (Mo) |
| | | Cobalt (Co) |

The first three of these elements -- carbon, hydrogen and oxygen -- are derived directly from the unlimited supplies of the atmosphere. Molecular nitrogen, although it makes up 78% of the air, has first to be changed into a simple compound (e.g., nitrate or ammonia) before it can serve the crop plants. This is accomplished by various types of bacteria, some that are associated with higher plants (legumes and alders), and some that live independently in the soil. Whether this *nitrogen fixation* is sufficient to ensure large crop yields depends on the nature of the soil, the supply of other nutrients, and especially on the amount of legumes in the crop rotation (the sequence of various crops in the same field).

Nitrogen, potassium and phosphorus are the elements most frequently limiting crop production and are therefore the three main fertilizer elements. In intensive farming, they must be added for top yields. Calcium, magnesium and sulfur are the secondary plant food elements. They also are needed in fairly large amounts by plants, but many soils contain sufficient quantities of these elements, or they are supplied as lime or as incidental ingredients of other fertilizers.

The elements listed in the third column of table 5-2 are the minor plant food elements, also called the micro-nutrients. The word *minor* refers merely to the amounts that are needed for the plants, not to their importance. The minor plant food elements are just as essential for plant growth as any of the other elements. Most of these nutrients are derived from the soil, but occasionally some are deficient and need to be added.

## Nutrient Levels and Crop Yields

For crop plants to give the highest possible yields, each of the growth factors must exist in optimum amounts. The physical growth factors have already been discussed. Plants require different amounts of the various nutrient elements. These amounts can be determined through experiments in which the level of only one element is varied and all other nutrients are well supplied. This is illustrated for potassium in figure 5-1, which shows that the first increments of available potassium give the greatest yield increases. As the maximum yield is approached, increases in the level of the nutrient bring about only small yield increases. *Maximum yield* is the yield a crop can achieve under the given physical and chemical conditions if potassium, or other essential nutrients, are available in optimum amounts. This *law of the growth factors* applies to each of the nutrient elements. For example, if there is only enough phosphate for half of the maximum yield and enough potassium for

half of the maximum yield, there will actually only be half
of half (or one-fourth) of the yield that could be obtained
if phosphate and potassium were available in sufficient
amounts.

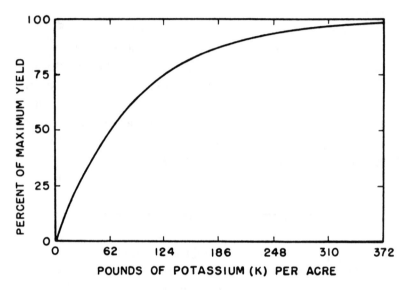

Fig. 5-1. The effect of potassium level in the soil on crop yield.
"Maximum yield" is the yield the crop can achieve under given
conditions of all other growth factors if potassium is available in
optimum amounts.

## Why Fertilize?

Under undisturbed, natural conditions any elements that
are taken up by the plants from the soil are returned to the
soil when the plants die and decompose. However, in
farming a sizable portion of the plant, sometimes the entire
above-ground part, is removed year after year, and eventually the store of nutrients in the soil is diminished to such
an extent that plant growth suffers. Besides, in their natural

condition the productive capacity of soils varies greatly. Some are so rich that only the plant food elements removed by the crops will have to be replaced to maintain large yields. Other soils are inherently poor and require large amounts of fertilizer to increase their yielding capacity. The first elements to become deficient are usually nitrogen, phosphorus and potassium. But insufficient amounts of any of the other elements supplied through the soil have caused seriously low yields in various locations.

## Determining the Needs for Fertilization

Commercial fertilizers are needed wherever intensive agriculture is practiced to correct plant nutrient deficiencies and to produce satisfactory yields. The problem is, which elements are required in a given field and how much should be applied?

This question resolves itself into two aspects: How much of the plant-available form of the element does the soil contain? Secondly, what are the specific nutrient requirements of the crop to be grown?

### Chemical Soil Tests

Numerous chemical tests for plant nutrients in soils have been devised. Most of these methods use a solution to extract the *available* nutrients and then determine the amounts of nutrients in this extract. Frequently, only phosphorus and potassium are determined. It is more difficult to test for available nitrogen, because nitrogen undergoes many chemical interchanges between organic and inorganic forms in the soil. Thus, the amount of available nitrogen changes during the growing season. However, some nitrogen tests are being used on samples collected at the same time during the growing season each year, usually in late spring. Available calcium and magnesium can be determined in the exchangeable form, similar to the potassium determination, but usually the need to use these elements is based on the soil reaction (pH). Soil tests have been developed for some minor elements, but they are not used as extensively as those for phosphorus and potassium. Because of the very low concentrations of micro-nutrients, special precautions must be taken to avoid contamination during sampling and analysis.

### *Biologic Soil Tests*

Even though tests for available phosphorus and potassium have been used for more than half a century, their results are not always as reliable as could be desired. Therefore, a number of biologic soil tests have been developed. It is best to use an indicator plant in soil fertilized at different levels to determine which of the main nutrient elements is deficient. This method is very dependable, so it is presented here in detail.

Soil is mixed with combinations of salts of nitrogen, phosphorus and potassium, placed in small containers, and several small seedling plants (e.g., lettuce or tomatoes) are planted in each of these containers. The soils are kept moist with distilled water. The growth of the plants is observed and their height measured. Any deficiency symptoms are noted. After four to six weeks the plants are cut off at the soil surface, dried and weighed. The treatments shown in table 5-3 are used.

Table 5-3. Research plan to determine fertilizer needs.

| Treatment | Fertilizer Elements Applied | | |
|:---:|:---:|:---:|:---:|
| 1 | N | P | K |
| 2 | N | P | - |
| 3 | N | - | K |
| 4 | - | P | K |
| 5 | - | - | - |

N = 0.1 g N (or 0.33 g $NH_4NO_3$) per 1000 g of soil.
P = 0.1 g P (or 0.41 g $Ca(H_2PO_4)_2 \cdot H_2O$) per 1000 g of soil.
K = 0.2 g K (or 0.24 g KCl) per 1000 g of soil.
- = No fertilizer applied.

From the yields of the five treatments, the need for each of the three main nutrients can be deduced. If, for instance, the yield from treatment 3 is much smaller than that from treatment 1, the soil is low in available phosphorus (P). This method can be modified to include several levels of the three nutrients and can thus give an idea of the amount of the fertilizer required.

## Plant Testing

The most direct method to make sure crops are receiving enough nutrient elements from the soil is to analyze the plants themselves. Two approaches can be used: foliar analysis and tissue testing.

In *foliar analysis* certain leaves of the plant are collected and sent to a laboratory where the elemental composition is determined. These results are then compared with standards previously established for the given crop.

Spectrographic determination of ten or more elements in plant material can be done quickly and inexpensively by laboratories set up for this service.

*Plant tissue testing* is best done directly in the field. Sap is pressed out from the plant tissue and its content of nitrate, phosphate, potassium and other elements determined semi-quantitatively. Test kits are available to test for three or four elements of one plant in a few minutes. Plant tissue testing can quickly tell whether the supply of a nutrient is "high," "medium" or "low." It is important to test the plants in the right stage of development and to choose the right part of the plant. Interpretation of the results of the tissue tests is frequently more difficult than conducting the tests themselves.

## Balance Sheet of Gains and Losses of Plant Nutrients

Soils gain fertility by material brought down in the rain, by nitrogen fixation, by decomposition of crop residues, by the slow weathering of the soil and, of course, by manuring and fertilizing. On the other hand, they lose fertility by removal of crops, by erosion, by leaching and by a change-over of nutrient elements from the available to the unavailable form. To maintain the fertility status of a soil, the gains must equal the losses; to increase the fertility status, the gains should be larger than the losses. Unfortunately, there are several factors in this equation that cannot be determined with a great deal of accuracy. However, a fairly sound estimate can be obtained from a balance sheet. This estimate plus a knowledge of the requirements of the next crop to be grown can form a useful basis for a fertilization plan.

The amounts of plant nutrients taken up by crops and removed from the field vary a great deal from soil to soil and from year to year (see table 5-4 for some approximate estimates). The nitrogen, phosphorus and potassium composition of fertilizers is expressed on their labels as percent nitrogen (N) - percent phosphorus pentoxide ($P_2O_5$)- percent potassium oxide ($K_2O$). This type of expression started when these chemicals were actually weighed in the laboratory. Although this method was replaced long ago by faster and more accurate techniques, the labels continue to use the oxides $P_2O_5$ and $K_2O$. Since these oxides occur neither in soil, fertilizer, nor plants, the elemental method of designating the composition (N, P and K) is now generally adopted for research reports. To calculate the elemental percentages, the conversion factor from percent $P_2O_5$ to percent phosphorus is 0.44, and that from percent $K_2O$ to percent potassium is 0.83. For example, a 10-10-10 fertilizer according to the label designation method (N- $P_2O_5$ - $K_2O$) would be a 10-4.4-8.3 fertilizer according to the elemental method (N - P - K). In table 5-4 the composition of the crops is expressed as N - $P_2O_5$ - $K_2O$ so that growers can calculate directly how much fertilizer must be added to replace the nutrients removed by crops.

Table 5-4. Amount of plant nutrients removed by crops.

| Crop | Yield per acre | N | $P_2O_5$ | $K_2O$ |
|------|------|------|------|------|
| | | --------pounds/acre--------- | | |
| Corn grain | 120 bu. | 90 | 44 | 35 |
| Corn silage | 15 tons* | 125 | 39 | 105 |
| Soybeans | 37 bu. | 148[+] | 33 | 52 |
| Wheat | 45 bu. | 68 | 25 | 15 |
| Alfalfa hay | 3 tons | 168[+] | 45 | 150 |

\* Wet weight
+ Legume crop fixes some N from the air.
From *Corn and Soybean Guide*, 1993, Crop Diagnostic Training and Research Center, Purdue University, West Lafayette, Indiana.

In general, various crops have different nutrient requirements. Legumes require much potassium and calcium. Corn requires much nitrogen, phosphorus and potassium. Wheat requires much phosphorus and fair amounts of nitrogen; it seldom starves for potassium or suffers from acidity. Crops requiring particularly high amounts of plant nutrients are alfalfa, sugar beets, tobacco, corn, tomatoes and other truck crops. Crops requiring relatively small amounts of plant nutrients are grasses, small grains and forest trees, especially pines.

## Yield Levels and Deficiency Symptoms

Yield levels are a helpful guide to determine the general fertility status. Where the yields are as high as the climate and the physical conditions of the soil permit, no major nutrient deficiencies are likely to exist. Unsatisfactory yields, however, are no indication of which of the nutrients are deficient, or even if any of the nutrients is deficient in the soil. Drought, excessive wetness, poor physical condition of the soil, and diseases are other factors that may have kept the yield down.

If the deficiency of certain plant nutrients is pronounced, plants exhibit signs of this specific starvation. Nitrogen deficiency turns the center of corn leaves yellow, especially the bottom leaves. Phosphorus deficiency stunts corn plants and frequently turns them purplish. Potassium deficiency results in "marginal firing" (yellowing of the rim) of corn leaves. Magnesium deficiency shows in parallel, light streaks on corn leaves. Similar hunger signs have been noted in practically every crop, but they usually only occur after the deficiencies have become serious. Obviously, visible deficiency symptoms and reduced crop yield cannot be used to correct soil fertility problems in the present year, but they are a good guide for future fertilization.

## The Essence of a Fertilization Program

The principles of fertilization for increased crop production are summarized here. Bring the soil into the best physical condition by use of adapted rotations, with as much of the land in grass-legume meadows as economical, and by judicious cultivation. Leave as much of the residue in the field as practical and return residue that has been used as bedding for livestock together with the manure back to the fields. Apply sufficient lime to bring the pH of the soil to about 6.5 and add enough fertilizer to make up for any losses due to crop removal and other causes.

The soil is the main working capital of the farmer. Remember that you have to replenish your bank account if you want to continue to write checks; a well-supplied account permits much greater financial flexibility and a better chance to make use of business opportunities than one that limps along near the zero mark.

Fertilization has to be viewed in light of the entire farm operation. Fertilizer should only be applied to land that has sufficient productive potential. Land that is too wet, too subject to drought, too acid, or has other physical factors limiting crop production should only be fertilized after these detrimental properties have been rectified. Over-fertilization may result in pollution of ground water by leaching through the soil or pollution of surface water from runoff and erosion. This condition is corrected by removing some of the fertilizer elements through the crops. Thus, over-fertilization may be a more serious problem than under-fertilization because it is more difficult to correct.

Once the plant nutrient status of the soil is known, it must be decided how large a crop can be reasonably expected under the given conditions of soil management and climate. The larger the potential productivity, the larger the amount of fertilization required. Economic considerations will determine whether the calculated amount should be used. If the soil is very low in plant nutrients, it may not be practical to bring it up to the desired status all in one year. In such a case ample row fertilization (for the row crops) is best. This will not give maximum yields but will give satisfactory yields economically. After a few years this type of fertilization can raise the nutrient level high enough that a broadcast fertilization can bring the entire body of the soil to the ideal level. Even on fairly rich soils, it is usually well to apply some row fertilizer. Many of the important row crops have only a limited ability to absorb phosphorus, and sometimes also nitrogen and potassium, while they are young. Row fertilizer is best placed only on one side of the row of plants and about an inch or two below seed level (fig. 5-2). This gives the roots a chance to obtain the necessary nutrients without suffering from too high a salt concentration in the soil solution.

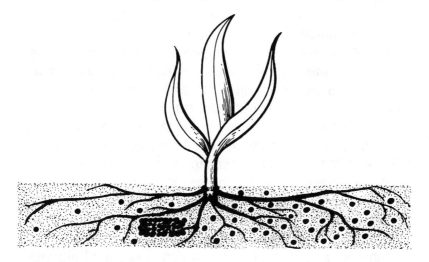

Fig. 5-2. A combination of row and broadcast fertilization assures the best growth.

### *Fertilizing with Nitrogen*

Plants take up nitrogen mostly as the nitrate ion ($NO_3^-$) and as the ammonium ion ($NH_4^+$). Biological reactions change the nitrogen compounds that occur in the soil to the ammonium and the nitrate forms. The ammonium ion is held by the clay, similarly to the potassium ion and the calcium ion, and therefore is not lost from the soil. The nitrate ion, however, is not held by the soil and moves with the water. In warm, moist, well-aerated soils, ammonium is quickly changed, first to nitrite and then to nitrate. In nitrate form nitrogen is liable to be washed out of the soil. In the summer the danger that this will occur is not great, since much of the rain is evaporated from the soil or used for transpiration by the plants. Another reason that nitrogen losses in the summer are generally not very serious is that the soil microbes use up the nitrate to form the organic compounds of their bodies, and in this way the nitrogen is not able to get away. In the winter much leaching occurs, and large amounts of nitrate are lost. It is therefore not advisable to apply more nitrogen fertilizer than the crop can use in one season. In fact, where much rain occurs during the summer or where the land is irrigated, nitrogen fertilizer is applied two or more times during the summer. As this is inconvenient and not always effective, attempts have been made to slow down the change-over to nitrate. Slowly soluble nitrogen compounds (e.g., urea formaldehyde) are used, or the nitrogen fertilizer is mixed with slowly soluble organic materials. Sometimes a chemical inhibitor is added to the soil that slows down the activity of the bacteria that convert ammonia into nitrite.

Anhydrous ammonia ($NH_3$) is a commonly used nitrogen fertilizer. At the same time its strongly alkaline reaction kills the nitrifying bacteria in the immediate vicinity of the applied ammonia and thus reduces the rate of change-over to nitrate. In this way the nitrogen remains in the form of ammonia and is retained by the soil because it is protected from leaching. As the pH is gradually lowered, microbes reenter the zone and begin to convert the ammonia to nitrate. It is the task of the farmer to time the ammonia application so that the nitrate becomes available when it is needed. Practically all soils require nitrogen fertilizers to give maximum crop yields. Nitrogen fertilizer rates are set to equal the amount expected to be removed by the current crop. If less nitrogen is applied, crop yield might be reduced. If more nitrogen is applied, the excess may leach into the ground water or surface water.

## Fertilizing with Phosphorus

Phosphorus is taken up by plants in much smaller amounts than nitrogen and potassium, yet frequently so little of it is present in the soil in a form available to plants that the plants suffer and the soil produces poor yields. Phosphorus is taken up mostly as dihydrogen phosphate ($H_2PO_4^-$) and to a small extent as monohydrogen phosphate ($HPO_4^{2-}$). When phosphorus combines in the soil with calcium or magnesium to form calcium phosphate or magnesium phosphate, as happens when the pH of the soil is 6.0 or higher, it remains fairly well available to the plant roots. In an acid soil, however, the fertilizer phosphorus soon changes to iron phosphate or aluminum phosphate. Both of these forms are practically worthless for plant growth. Phosphorus is also tightly absorbed by the clay. On acid soils, it is best not to broadcast the phosphorus fertilizer and to mix it thoroughly with the soil but to apply it in bands or as pellets so that it does not suffer too great a loss of availability. To make phosphate application effective, it is well to first lime the soil to a pH of at least 6.2 and to use it in combination with nitrogen or an organic fertilizer. As long as the level of available phosphate is low, band or hill placement of phosphorus fertilizer is preferable to broadcast application. Under any conditions only 10 to 15% of the applied phosphorus is taken up by the crops in the first season.

When the soil test shows that phosphorus is needed, more phosphorus has to be applied than the amount missing from the desired level. As an example, if the test shows 40 pounds of available phosphorus per acre and the level desired for the next crop is 100 pounds per acre, it is necessary to apply considerably more than 60 pounds of fertilizer phosphorus. The reason for this is that part of the fertilizer phosphorus is fixed in unavailable form, at least for the time being.

## Fertilizing with Potassium

Potassium fertilization presents only a few problems. It neither changes over to a form that is readily washed out with the drainage water -- as is the case with nitrate -- nor is it held as tenaciously as phosphate. The main potassium fertilizer is potassium chloride (KCl). If this fertilizer is placed very near a seed or a root, the high osmotic concentration created in the soil solution can be injurious to plants. If large amounts of potassium are needed, it is best to broadcast potassium chloride several weeks or months prior to the planting of the crop. Thus, high salt concentrations are avoided and some of the chloride is leached out because it is not held tightly by the soil. Potassium is absorbed by the negative sites on clay and organic matter fractions of the soil and can be exchanged for other cations in the soil solution or for cations given off by plant roots. In this way it is available to plants. If potassium fertilizer is placed on top of the ground, practically all of it stays in the surface inch of the soil and is used by the roots in this layer. Smaller amounts can be placed in bands with the nitrogen and phosphorus to start the crop growing.

One of the crops that needs much potassium is alfalfa. To maintain a stand of alfalfa for several years, the soil should be well supplied with potassium before seeding and more potassium must be broadcast periodically.

### Secondary and Minor Elements

Calcium and magnesium are usually supplied in sufficient quantities where liming to a pH level of around 6.5 is practiced. If the magnesium level is found to be low, liming should be done with dolomitic limestone.

Formerly, sulfur was supplied in superphosphate and ammonium sulfate fertilizers. These fertilizers have been largely replaced by other phosphate and nitrogen compounds. Today, only small amounts of sulfur are added to the soil as incidental ingredients of other fertilizers. If it is found wanting, sulfur may be applied as gypsum ($CaSO_4$) or, in soils of high pH, as elemental sulfur.

Wherever minor elements are needed they can either be mixed with other fertilizer materials or they can be applied dissolved in water as a spray on the soil or directly to the plants. Great care has to be taken not to use too much, as even a small excess of some of the elements can be toxic to the plants.

### Liming

The purpose of liming is fourfold: to supply calcium (and sometimes magnesium) to the plants, to neutralize soil acidity (to raise pH), to help make other nutrient elements more available to plants, and to help create an environment optimal for soil microorganisms.

It is generally recommended that acid soils be limed to a pH of 6.0 to 6.5. At this soil reaction most of the plant nutrient elements are more readily soluble than at considerably more acid or more alkaline reactions. Also, favorable microbes are most active at this reaction. Soil pH can be determined in the field with liquid indicator dyes or with paper or plastic strips on which the dyes are absorbed. Soil testing laboratories can determine soil pH more accurately and can also perform tests on which lime requirement is based.

Because calcium carbonate ($CaCO_3$) and magnesium carbonate ($MgCO_3$), the essential ingredients of limestone, are only slightly soluble, it is important that lime be applied as a fine powder and that it be distributed throughout the soil. To be useful limestone should be ground sufficiently fine that at least 40% of it can pass through a 60-mesh screen (about 0.2 mm openings).

# 6

# ORGANIC MATTER AND
# MICROBES OF THE SOIL

## Classification of Organic Matter

All organic substances in the soil, living or dead, fresh or decomposed, are part of the soil organic matter. This includes plant roots, small animals, plant and animal residues, humus and microbes. The nature of the plant-root system has a significant effect on how organic matter is added to the soil and on how roots affect their environment. Tree roots are generally very coarse and they persist in the soil for many years. Under trees, organic matter is added on the surface of the soil. Grasses, on the other hand, have a fibrous root system, much of which dies every year and adds organic matter to the soil, mainly in surface and upper subsurface horizons. The tap roots of many legumes have the ability to penetrate deeply into the soil. Non-living organic matter can best be classified into two categories: humus and plant residues.

## Composition of Humus

*Humus* is the dark-colored soil organic matter. It has fairly definite chemical and physical properties and is not subject to as rapid decomposition as plant residues. Humus is a colloidal (glue-like) substance containing about 50% carbon, 5% nitrogen and 0.5% phosphorus. Chemically, it is a combination of modified lignin (the most decay-resistant constituent of the mature cell walls of plants) and amino acids (the components of proteins) and of other nitrogenous compounds. Humus represents the decomposition products of organic residue and materials synthesized by microorganisms.

## Factors Affecting the Formation of Humus

The amount and type of humus depends both on the amount and types of plant residues available to microorganisms and on the conditions under which the humus was formed. Moist and cool climates favor humus formation; dry and hot climates do not. Prairie soils are much richer in humus than soils that formed under forest vegetation. Forested soils have a considerable amount of humus near the surface but very little in the subsoil, while in the prairie soils the humus reaches to a depth of one or even two feet. This penetration is a result of the decay of the prairie grass roots.

A cultivated soil usually has a smaller humus content than it had in its pristine condition. This is due to removal of plant materials, to the increased oxidation as a result of tillage, and to erosion.

Because a high content of organic matter is of value to soil fertility, soil structure and erosion control, it must be the aim of every farmer to increase the organic matter content of his or her land to a point where these benefits will accrue. This is done by producing much plant material and by permitting it to decompose at a leisurely rate. Toward the first end, ample fertilization is practiced, adapted crops (including legumes and grasses) are selected, and most of the crop residue is left in the field or is returned in the form of manure. To facilitate the second aim, such extremes as burning, excessive cultivation and burying the residues on the bottom of a wet, sticky plow furrow must be avoided.

## Functions of Organic Matter in the Soil

Organic matter has the ability to combine with the mineral particles to form water-stable aggregates. It helps to make a tough clay soil mellow, and it imparts cohesion to a loose sandy soil. As a matter of fact, organic matter improves the structure and the water regime of any mineral soil. Together with this goes the improved aeration.

In its decomposition, organic matter breaks down mainly into minerals, water and carbon dioxide. This decomposition is most rapid in the warm part of the year when the crops require these minerals as nutrients. This constant supply of plant nutrients at the time of greatest need is an important function of organic matter. The carbon dioxide evolved helps to dissolve the soil minerals and to bring them into a form available to plant roots.

## Microbes in the Soil

Every productive soil teems with a multitude of microbes of the most diverse kinds. Microscopic plants such as bacteria, fungi, actinomycetes and algae, and animals such as nematodes and protozoa are abundant in the soil. All these organisms are instrumental in decomposing organic matter and releasing available plant nutrient elements.

Soil organisms are especially important for nitrogen (N) transformations in the soil, because they incorporate nitrogen into their tissues. Also, since nitrogen can exist in different oxidation states in the soil (table 6-1), soil organisms are responsible for transformations from one state to another.

Table 6-1.    Forms and oxidation states of nitrogen compounds in soils.

| Ion or molecule | | Oxidation state | Process |
|---|---|---|---|
| Name | Formula | | |
| Oxidized forms: | | | |
| Nitrate ion | $NO_3^-$ | +5 | Reduction |
| Nitrite ion | $NO_2^-$ | +3 | |
| Nitric oxide gas | NO | +2 | |
| Nitrous oxide gas | $N_2O$ | +1 | |
| Intermediate form: | | | |
| Nitrogen gas | $N_2$ | 0 | |
| Reduced forms: | | | |
| Ammonium ion | $NH_4^+$ | -3 | |
| Ammonium gas | $NH_3$ | -3 | Oxidation |

One kind of transformation is between organic and inorganic forms of nitrogen. *Immobilization* is the process by which inorganic forms of nitrogen, such as nitrate ($NO_3^-$) and ammonium ($NH_4^+$), are incorporated into the tissue of an organism. Immobilization occurs when a plant takes up these ions and forms nitrogen-containing organic compounds in its tissue. It also occurs when soil microorganisms absorb nitrate and ammonium, ingest low-nitrogen materials, and synthesize nitrogen-rich materials. *Mineralization* takes place when the plant or microorganism dies and is the process by which the organic compounds are broken down and nitrate and ammonium are released.

Soil organisms are also responsible for other chemical transformations of nitrogen. *Nitrification* is the oxidation of ammonia ($NH_3$) or ammonium to nitrate. It takes place in well-aerated soils. *Denitrification* is the reduction of nitrate to gases such as nitric oxide, nitrous oxide or elemental nitrogen. This process takes place in water-logged soils where organisms are unable to obtain oxygen. Both nitrification and denitrification are biological, so they do not occur when the soil is cold (less than about 5°C) and when the organisms are inactive.

Also of importance are the *nitrogen-fixing bacteria,* which have the ability to change the elemental nitrogen of the air into chemical compounds that plants can use. Some of these nitrogen fixers are active only in nodules attached to plant roots (especially legumes); others live freely in soil. Many other groups of microbes have similar specific functions, others merely help in the decomposition of plant material. They not only prepare the material locked up in the dead plant parts for further use, but they also help to create a good soil structure by excreting gelatinous materials which serve to bind the soil minerals together. Some of the microscopic animals are credited with actually creating humus in their digestive tracts.

No soil is without microbes: the more fertile a soil is, the more microbes it contains and the more active these organisms are. It is said that the carrying capacity of a pasture is equal to the weight of the microbes in its soil. The requirements of microbes are very similar to those of the higher plants. They need moisture, air, a favorable temperature range, and about the same plant nutrient elements. However, as microbes (at least most of them) cannot create organic matter through photosynthesis, they depend on the residues of other plants and animals for a source of energy. Striving for a high number of active microbes is, therefore, identical with striving for a soil condition favorable for crop plants.

When fresh plant residue is incorporated into the soil,

much readily available energy material of carbohydrates (sugars, starches, etc.) is utilized by bacteria. Their increased activity and growth results in high demands on the plant nutrients, especially nitrogen. Much of this is quickly tied up in compounds of the bodies of the bacteria and a temporary shortage is created in the soil. Crop plants respond to such a condition with slow growth and a pale color (indication of hunger for nitrogen). After the supply of carbohydrates is exhausted, the bacteria die and release minerals and nitrogen, and the crops prosper once more. A wise farmer adjusts the incorporation of organic matter in such a way that its most active decomposition does not coincide with the maximum nutrient requirements of the crops, or, he or she supplies some extra nitrogen in the form of commercial fertilizer when incorporating carbonaceous (low-nitrogen) residues into the soil. Where the ratio of carbon to nitrogen in the crop residues is 30:1 or less, no serious tie-up of the soil nitrogen occurs.

## Small Animals in the Soil

The small animals living in the soil that can readily be observed with the naked eye, include insects, spiders, mites, springtails and earthworms. Of these, the earthworm has gotten the most publicity, and probably with justification. Earthworms dig countless tunnels that help to distribute rain water and aerate the soil. They mix the materials of the different horizons, they transform plant and animal waste into rich humus, and they help to increase the amount of available plant food in the soil. The earthworms are truly valuable cooperators of the farmer. However, there are areas in the world where no earthworms exist, yet they are fertile and productive. We are apt to give credit to earthworms for soil improvements that may be due to some of the microscopic animals (mites, spiders, springtails, etc.) because a casual observer does not notice these tiny creatures. Conditions favorable for earthworms are also favorable for bacteria and for the microscopic animals, and it is difficult to distinguish between the results of the activities of these three groups.

# FACTORS AND PROCESSES
# OF SOIL FORMATION

Soils differ greatly from one part of the world to another and even locally, from the top of a hill to the bottom. If we are to care for our soils so they will last forever, we should know how they formed and how they relate to their environment. Soil properties depend on five factors. These *soil formation factors* determine the kind of soil that forms, but they do not explain how it forms. *Soil forming processes* explain the mechanisms by which soils form.

## Soil Formation Factors

The five factors of soil formation are *climate, organisms, relief, parent material,* and *time.* They can be remembered by thinking of the expression "clorpt." Two factors, parent material and relief (topography), describe the conditions when soil formation began. This system is acted upon by the "active" factors, climate and organisms, during a certain length of time.

### *Parent Material*

Parent material is the geologic material from which soils form. Below are listed some important parent materials.

| | |
|---|---|
| Igneous rocks | rocks that crystallized from molten lava, such as granite, diorite and basalt |
| Sedimentary rocks | rocks formed by sedimentation from water, such as sandstone, shale and limestone |

| | |
|---|---|
| Metamorphic rocks | rocks formed by alteration of igneous or sedimentary rocks under conditions of high temperature and pressure, such as gneiss, schist and marble |
| Glacial deposits | material carried by ice, such as glacial till, or by ice and water, such as outwash |
| Water deposits | alluvium (recent), old alluvium (stream), lacustrine (lake), coastal plain and beach deposits |
| Wind deposits | loess (silt-size), dune sand, and volcanic ash |

The kind of parent material may impose limits on the properties of the soil that forms from it. If a certain kind of rock is especially low in calcium, for example, the soil that forms from it will always also be low in calcium. Furthermore, the soil might inherit some important physical properties from the parent material. Glacial till deposits are often very dense and compact because of the weight of the ice that overrode the till when it was wet. Soil horizons forming in the till inherit this high density. In contrast, wind-blown loess is laid down quite loosely and the soils that form from it inherit this property. As a result, roots and water can move through soils formed from loess much more readily than they can move through soils formed from glacial till. Also, soil material, such as dune sand, that was deposited by wind is prone to being picked up by the wind again, so the soils which form from it are very subject to wind erosion.

## *Relief*

Relief, or topography, refers to the shape of the land surface when soil formation began. This factor greatly determines how water moves in the landscape and how susceptible the soils are to erosion by water. Water runs off steep slopes and ponds in depressions in the landscape. Steep slopes are very erodible and erodibility largely determines whether or not a soil can be farmed with ordinary farm implements. Also included in the relief factor is the shape of the water table surface, or the depth to the water table. Many flat landscapes have layers through which water moves very slowly, and these landscapes often have water-logged soils. The relief factor is very significant to crop production and to many other soil uses.

## *Climate*

The climate of a region greatly influences the kind and rate of soil forming processes that occur. Soil temperature influences the rate of chemical and biological reactions -- minerals weather much faster in warm than in cold climates. Temperature also influences the rate of water evaporation from the soil and the rate of transpiration by plants.

Precipitation is the other major component of the climate factor. Where it is high, soluble materials will be leached from the soil; where it is low, soluble materials will accumulate in the soil or on the surface.

## *Organisms*

This factor refers to plants, animals and microorganisms living on and in the soil. The kind of vegetation under which the soil formed influences some important soil properties, largely through the manner in which organic matter is added to the soil. Under forest vegetation, leaves fall and accumulate on the soil surface. These soils usually have a layer of organic matter on top of the mineral soil surface. Prairie grasses, on the other hand, have a fibrous root system in upper soil layers. When these plants die, organic matter is added throughout upper soil horizons, so the soil has a thick, organic-rich surface horizon (fig. 7-1).

Fig. 7-1. The two kinds of soil horizons that form under prairie and forest vegetation and intergrades between the two. The significance of the horizon designations at both ends of the diagram is explained in chapter 8. (From *Understanding and Judging Indiana Soils*, p. 8, by D. P. Franzmeier, J. E. Yahner, G. C. Steinhardt, and D. G. Schulze, 1989, Cooperative Extension Service Publication ID-72, Purdue University, West Lafayette, Indiana.)

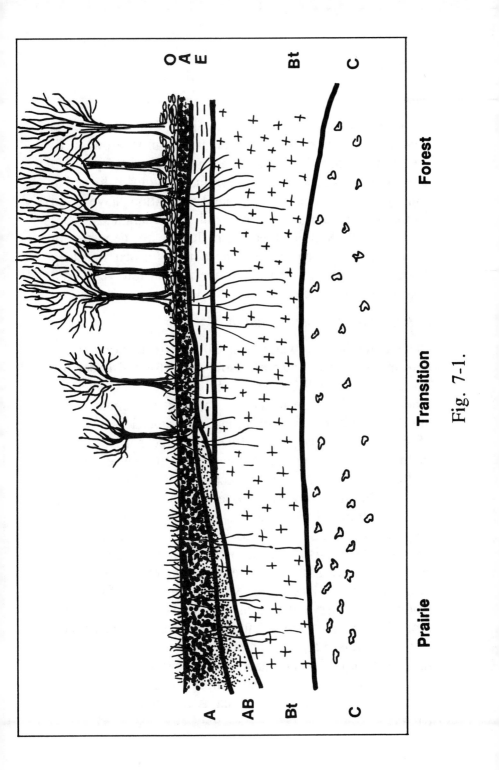

Fig. 7-1.

## Time

The time of soil formation is measured from the time the land surface was exposed and plants began to grow on it, not from the time when the parent material was deposited. The time of soil formation determines how long the active soil formation factors, climate and vegetation, have acted on the parent material to form a soil. Some upland surfaces are very old. Soils formed from flood plain deposits are usually quite young, as are those that formed on steep slopes. Soil landscapes vary in age from a few years to a few million years old.

## Significance of Soil Formation Factors

The conditions or values for all soil formation factors determine the properties of a soil. Soils formed under the same conditions will have the same properties regardless of where the soils were formed. Changing the conditions of any factor will result in different soil properties.

Soil scientists study the influence of the soil formation factors by observing the change in soil properties as one factor changes and others are kept constant. For example, they might study how soil properties change from the top to the bottom of a hillslope when the parent material, climate, vegetation and age of land surface are held constant. They might also study how soil properties change as rainfall decreases from eastern to western United States. Soil scientists deduce the effect of time on soil formation when they observe that soils on old land surfaces have more distinctive horizons than those on recent surfaces such as flood plains. Large geographic areas are needed to study the effects of regional climate and vegetation differences on soil properties. For this reason, these kinds of studies were first done in large countries such as the United States and Russia.

## Soil Forming Processes

Processes of soil formation include the physical, chemical and biological processes by which distinctive layers, called soil horizons, form in the upper meter or two of the parent material. These processes describe the mechanisms, actions or changes that take place as the soil forms from its parent material. Some of the main soil forming processes are described in this section.

### Physical Weathering

Larger rocks and minerals are broken into smaller units by processes such as wetting and drying, freezing and thawing, abrasion as by the action of streams and glaciers, and by other physical forces. These processes break the rock into smaller particles that are more subject to chemical weathering.

### Chemical Weathering

Chemical weathering is one of the main processes of soil formation. Silicate minerals consist mainly of oxygen, silicon, aluminum, iron, magnesium, calcium, potassium, sodium and hydrogen. Through chemical weathering these minerals are transformed into insoluble materials that remain in the soil and soluble components that may be used by plants or leached from the soil. The new solid materials can be formed by a minor alteration of the parent mineral structure (such as loss of potassium ions), by weathering to small structural remnants and a recombination of them, or by dissolution to ions (or soluble molecules) and re-precipitation into insoluble materials. Mineral weathering is accelerated by acid soil conditions -- high hydrogen ion ($H^+$) concentration. The hydrogen ion is small and can invade mineral structures and replace other cations such as the potassium ion ($K^+$), the sodium ion ($Na^+$), the calcium ion ($Ca^{2+}$), and the magnesium ion ($Mg^{2+}$). With this replacement the structures become unstable. These unstable structures can break into silicon-aluminum remnants, which can form clay minerals, or they can break down all the way to their component ions.

## *Leaching*

Leaching is the removal of materials in solution from a soil horizon. Cations such as $Na^+$, $K^+$, $Ca^{2+}$ and $Mg^{2+}$ are released to the soil solution during mineral weathering. Some are adsorbed as exchangeable cations, but many are leached from the horizon, that is, they remain in the solution as it moves down the profile. When the cations leach, they are accompanied by anions such as chloride ($Cl^-$), sulfate $SO_4^{2-}$, hydroxide ($OH^-$), carbonate ($CO_3^{2-}$), and bicarbonate ($HCO_3^-$).

Silica ($H_4SiO_4^0$), a molecule with no charge, is the soluble form of silicon in the soil solution. It is less soluble than compounds composed of the cations and anions described in the paragraph above, such as $NaCl$ or $CaCO_3$, but more soluble than aluminum or iron oxides, $Al_2O_3$ or $Fe_2O_3$. Therefore, silica tends to be leached from horizons. It can move to lower horizons where it can precipitate by combining with other ions or by evapotranspiration of water from the soil solution, or it can be leached from the soil entirely. In old, freely drained, highly weathered upland soils, most of the silicon present in the parent material has been leached, and the clay minerals formed in the soil contain very little silicon. On the other hand, soils in depressions low in the landscape accumulate silicon, and the clay minerals in those soils tend to be high in silicon.

If a chemical weathering reaction is to continue, the products formed in the reaction must be removed from the system. A common mineral in the soil is potassium feldspar, or orthoclase. In acid soil solutions (rich in $H^+$) it dissolves to release potassium ions ($K^+$), aluminum ions ($Al^{+3}$), and silica according to the reaction:

$$KAlSi_3O_8 \; + \; 4\,H^+ + \; 4\,H_2O \; \rightleftharpoons \; K^+ \; + \; Al^{3+} + \; 3\,H_4SiO_4^0$$

| Orthoclase | Hydrogen ion | Water | Potassium ion | Aluminum ion | Silica |

The equality symbol, $\rightleftharpoons$, indicates that the reaction can proceed from left to right or vice versa. If the reaction is to proceed to the right, the reaction products (those on the right side of the equation) must be removed from solution. If they are not, the rate of the reaction to the right equals the rate to the left and no net change results. Since aluminum ions readily combine with other ions in solution and precipitate to form insoluble materials, they are thus removed from solution. Potassium ions can be absorbed by clay minerals, and are thus removed from solution or leached from the soil. The main factor that controls the rate of the reaction is the fate of silica. If it is removed by leaching, the reaction proceeds to the right. If it is not removed, the reaction stalls.

The effects of the relative forward and backward rates of the reaction are observed in the field. Freely drained soils high in the landscape have relatively low feldspar contents and low-Si clay minerals. Soils in the depressions contain more feldspar and have high-Si clay minerals, formed from the silica that leached downslope.

The overall progress of this reaction depends on four main factors: (1) the resistance of the mineral to weathering, (2) the conditions that determine the rate of the reaction such as temperature, (3) the rate at which the products are leached from the horizon, and (4) the length of time the reaction proceeds.

## Accumulation of Organic Matter

Most parent materials are devoid of organic matter, but as soon as plants begin to grow on the developing soil, organic matter begins to accumulate in it, especially in surface horizons. As previously mentioned, organic matter accumulates differently in forest soils than in prairie soils. In the forest, organic matter is added mainly as leaf fall *on* the surface. Grasses, however, have a fibrous root system, and every year some of these roots die, resulting in addition of organic matter *in* the soil. Thus, forest soils have a thin, dark horizon over a leached horizon, while prairie soils have a thick, dark horizon (fig. 7-1).

## Clay Movement

One of the processes that leaves a distinctive mark on soil morphology is movement of clay particles from upper (E) horizons to lower (B) horizons. Some of the clay that accumulates in the B horizons form coatings, called *clay skins*, on soil peds or in pores. Clay skins are significant to the dynamics of the soil, because water moves and roots grow through voids in the soil lined with clay skins. The presence of clay skins is used in applying designations to soil horizons and in classifying soils.

## Complexation of Iron and Aluminum

In some soils, organic compounds combine with iron (Fe) and aluminum (Al) to form a soluble complex that moves from upper to lower horizons. This process is important in sandy soils and results in a distinctive morphology with light gray upper horizons, from which iron and organic matter have been removed, and reddish brown lower horizons where the Fe-organic matter complex accumulates.

## Oxidation-Reduction

In subsoil horizons, soil color is due mainly to iron oxide minerals, or to the lack of these minerals. The brownish or reddish color of well-aerated soils is due to iron oxide minerals that coat grayish minerals such as quartz or feldspar. In waterlogged soils, however, there are few iron oxides, so the soil color is due mainly to the grayish minerals.

In the process of respiration, soil microorganisms produce electrons as they consume food, mostly soil organic matter. In well-aerated soils, the *aerobes,* one set of microorganisms, utilize the organic matter they feed on and oxygen from the air, and they give off carbon dioxide. In this process electrons are transferred from carbon (C) of organic matter to oxygen in the soil air to form carbon dioxide.

Oxygen moves very slowly through water, however, so when the soil becomes water-logged, oxygen is not available to the organisms that require it. Then another set of microorganisms, the *anaerobes*, takes over. The anaerobes transfer electrons from C of organic matter to other atoms, such as nitrogen, iron (Fe), and manganese, and we say that C is oxidized while the other atoms are reduced.

Students remember the convention by using the mnemonic aid *oil rig*--oxidation **is loss** and reduction **is gain** in electrons. The reduction of hematite, an iron oxide mineral, is represented by the equation

$$Fe_2O_3 \;+\; CH_2O \;+\; 7\,CO_2 + 3\,H_2O \;\longrightarrow\; 4\,Fe^{2+} \;+\; 8\,HCO_3^-$$

| Hematite | Organic matter | Carbon dioxide | Water | Ferrous ion | Bicarbonate ion |
|---|---|---|---|---|---|

This chemical equation, which takes place in a saturated soil, illustrates that hematite (a red soil mineral) reacts with organic matter, carbon dioxide and water in the soil to produce soluble reduced iron (ferrous) and bicarbonate ions of the soil solution. In this reaction, C of organic matter loses electrons and Fe of hematite gains electrons. This process has important consequences for soil morphology, especially the soil color pattern. Iron oxide minerals such as hematite and goethite (a brownish mineral) are very insoluble. The reduced form of Fe ($Fe^{2+}$), however, is highly soluble. When a soil becomes waterlogged, $Fe^{2+}$ moves readily with the soil solution. It might move a short distance within a soil horizon, or entirely out of the soil. If it moves a short distance it can re-precipitate to form an accumulation of iron oxide minerals when the soil drains and becomes well aerated and oxidized. This gives a *mottled pattern* with gray zones of Fe depletion and brownish zones of Fe accumulation. If $Fe^{2+}$ moves entirely out of the soil horizon, it appears gray, the color of minerals like quartz and feldspar.

Manganese (Mn) is another element that can be oxidized and reduced in soils. In its oxidized form it is black; in its reduced form it is soluble and mobile in the soil.

## Cementation

Cementation occurs when some of the weathering products are leached from one horizon and accumulate in a lower one, where they may precipitate and cohere the soil particles to form a pan. Horizons can be cemented by calcium carbonate (lime), gypsum, silica or iron oxides.

## Shrink-Swell

Certain kinds of clay minerals swell when they get wet and shrink as they dry. If a soil with a high content of these minerals is in a climate that has distinct wet and dry periods, the whole soil will swell and shrink. When the clay minerals are dry, deep cracks form in the soil. When rains resume, the soil swells, the cracks close, and great internal forces are produced within the soil. These forces push the soil up in some places to form small knolls, leaving depressions between the knolls. The resulting landscape is said to have *gilgai* relief. Swelling soils present special problems for farming, and building roads and houses.

## Salt Accumulation

In arid areas where evapotranspiration exceeds precipitation, soluble salts are not leached out of the soil. The salts are composed mainly of the cations $Ca^{2+}$, $Mg^{2+}$, $Na^+$, and $K^+$, and the anions $Cl^-$, $SO_4^{2-}$, and $CO_3^{2-}$ or $HCO_3^-$. In fact, salts might move upward from the ground water into the soil, and they can accumulate to the point where they retard plant growth. When the soil dries, the salts may precipitate on the surface or in subsurface horizons.

## *Classifying Soil Horizons*

The soil forming processes discussed in this section produce distinctive layers in the soil. These layers, or soil horizons, are assigned certain horizon designations (see chapter 8). Some of these horizons qualify as diagnostic horizons, as explained in chapter 9.

# 8

# SOIL MORPHOLOGY

Soil scientists have developed methods to describe soils in the field so they can transfer information from one person to another and from one place to another. The methods used to describe *soil morphology,* the observable soil physical properties, has become well formalized.

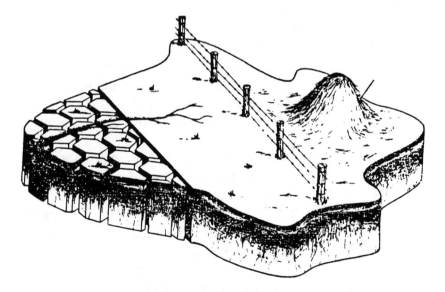

Fig. 8-1. A delineation on a soil map with a small part on the left divided into pedons. (From *The Soil Series in the U.S.A.*, Volume 2, pp. 17-24, by R. W. Simonson, 1964, The 8th International Congress of Soil Science, Bucharest. Reprinted by permission of the publisher.)

The basic unit of soil in the field is called a *pedon* (fig. 8-1). It is a three-dimensional body about one meter square by 1.5 to 2 meters deep. Soil scientists make detailed morphological descriptions of pedons. First, they pick out different layers, or soil horizons. They do this using properties they can see, feel, or otherwise identify, such as noticing how difficult it is to dig into the soil. For every horizon, they record the depth, assign an alphanumeric horizon designation, and describe several properties.

## Horizon Designations

Horizon designations include several components, as illustrated in figure 8-2.

- The number before the upper case letter represents the kind of parent material, counting from the top. The symbol 1 is not used; if no numeric symbol appears before the upper case letter, parent material 1 is assumed. This indicates that many soils formed in just one material.
- An upper case letter (one for every horizon) represents the major kind of soil horizon. The letters O, A, E, B, C and R are used, as explained in table 8-1.
- Following this letter, there might be a lower case letter providing more information about the nature of the horizon, as explained in table 8-2.
- Next, there could be a number that represents a subdivision of a horizon into two similar subhorizons.

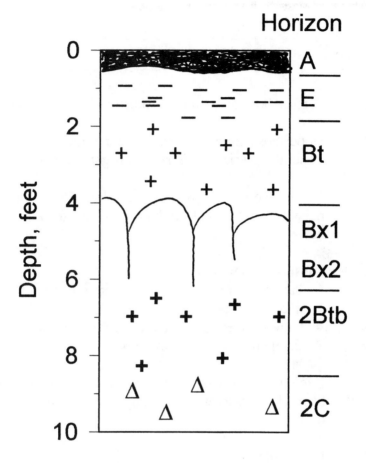

Figure 8-2. Sketch of a soil profile--a two-dimensional view of a pedon that shows various kinds of horizons and their designations.

Table 8-1. Master soil horizon designations.

| Symbol | Brief Description |
| --- | --- |
| O | Surface horizon composed mostly of organic matter |
| A | Uppermost horizon, which is composed mostly of mineral material but is rich in organic matter |
| E | Subsurface horizon from which clay, Fe and/or Al have been removed; lighter in color than the A horizon |
| B | Subsurface horizon that:<br>• has an accumulation of clay, organic matter, Fe and/or Al, or<br>• shows other evidence of alteration, such as distinctive color or structure |
| C | Geologic materials, including partially weathered bedrock, that show little change due to soil formation |
| R | Hard bedrock |

Table 8-2. Some subordinate horizon distinctions.

| Symbol | Brief Description |
| --- | --- |
| g | Strong gleying (gray colors); indicates reduced (wet) soil conditions |
| k | Accumulation of carbonates (lime) |
| o | Accumulation of Fe and Al as oxides, with little organic matter accumulation |
| p | An A horizon that has been plowed or otherwise tilled |
| s | Accumulation of Fe, Al and organic matter |
| t | Accumulation of clay |
| w | Weak B horizon development |
| x | Fragipan (weakly cemented subsoil horizon) |

In the example in figure 8-2, the soil has a thin A horizon that is enriched with organic matter. Below it is an E horizon from which clay has been removed, to be deposited in the Bt horizon. Below the Bt horizon is a fragipan that is divided into two subhorizons for sampling, the Bx1 and Bx2 horizons. All these horizons formed in loess, 6.5 feet deep, which buried the Bt horizon of an older soil that had been eroded, now the 2Btb horizon. This horizon was formed from glacial till, like the 2C horizon below it.

## Soil Morphology

Several properties of each horizon constitute a soil morphological description. The nature and significance of these properties are discussed below.

### *Color*

Color is described by comparing the color of a soil horizon with a standard Munsell color chart that looks something like a paint color chart. Often a soil horizon has several colors, and all are described. In surface horizons, soil color is related to the organic matter content of the soil. Dark-colored horizons contain more organic matter than light-colored ones. In subsurface horizons, soil color represents the water and aeration status of the soil. In general, a soil with a uniformly brownish or reddish subsoil is not saturated with water very often and thus has good aeration. A soil with mixed brownish and gray colors (mottled color pattern) has a water table that fluctuates often during the year and has brief periods when oxygen is in short supply. A soil that has a uniformly gray color has probably been saturated for a few months each year when the soil is warm. Gray soil colors are due to chemical reduction processes, as discussed in chapter 7.

### *Texture*

Soil texture refers to the particle-size distribution of the soil. Texture classes are illustrated in the triangular diagram in figure 2-4 on page 9. Texture affects many other soil properties, such as the ability to retain plant nutrients, the capacity to hold water for plant growth, and the rate of water movement through the soil. These properties are all very important for plant growth and for other uses of a soil.

Soil scientists estimate texture in the field by moistening a small sample of soil and manipulating it in their hand and feeling it. Sand particles feel gritty, silt particles feel floury, and clay particles feel sticky. They squeeze the soil to see how well it holds together. For finer-textured soils, they knead the soil to the consistence of modeling clay or putty and then squeeze it between their fingers to form a ribbon. The longer the ribbon, the more clay in the soil. Soil scientists and students calibrate their field estimates using samples for which particle-size distribution has been determined in the laboratory.

## *Structure*

Soil structure refers to the manner in which individual particles are held together to form aggregates, or peds, in the soil. Soil structure also determines the kind of space between soil particles or soil peds, especially the larger pore space through which water and air move readily. Three aspects of structure are described: the *size* of the peds, the *shape* of the peds and how *distinctly* they are formed. Shape is illustrated in figure 8-3. Size classes vary from a few millimeters to tens of centimeters. Distinctly formed peds often have surfaces coated with material somewhat different from the material inside them. These peds are very visible in a pedon, and they fall out when digging in the soil, especially when it is dry. Weakly formed peds are barely discernible.

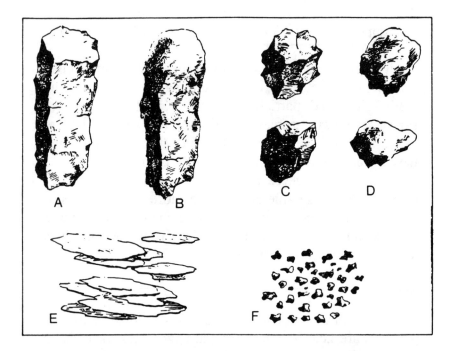

Fig. 8-3. Various shapes of soil structural units, or peds: A, prismatic; B, columnar; C, angular blocky; D, subangular blocky; E, platy; and F, granular. (From *Soil Survey Manual*, p. 227, U.S. Department of Agriculture Handbook no. 18, 1951, Washington, DC: U.S. Government Printing Office.)

## *Consistence*

Consistence refers to the strength of the soil. It varies from *loose*, like in a sand dune, to *cemented*, a soil horizon so hard that it cannot be dug out by hand. Soil consistence is very dependent on moisture content. It is estimated by how difficult it is to break a piece of soil -- that is, whether it can be broken with the fingers, between both hands, or by stepping on it.

## *Reaction*

Reaction is a measure of the relative acidity or alkalinity of the soil according to the pH scale, in which pH 7 is neutral, pH values less than 7 are acid, and values greater than 7 are alkaline. Soil pH values range mostly between 5 and 8.5. Horizons that have pH values less than 5 are likely to be so acid that aluminum compounds may be toxic to plant roots. Soil pH values greater than 8.5 usually are due to sodium; high sodium soils often have adverse physical and chemical properties. Also, at high pH values some nutrient elements become unavailable to plants. The presence of carbonate minerals (lime) is also noted in soil descriptions. Soil scientists test soils in the field for carbonate minerals by dropping a weak solution of hydrochloric acid (HCl) on them. If carbonates are present, escaping carbon dioxide gas ($CO_2$) can be seen bubbling through the liquid.

## *Other Horizon Properties*

Soil scientists also determine the shape and distinctness of the boundary between adjacent horizons. They describe the nature of ped surfaces, which is significant because water moves mainly along these surfaces and plant roots grow on them. They may also describe the nature of the visible pores and the size and abundance of roots in each horizon.

## Use of Soil Morphological Information

Pedon information serves as the basis for mapping soils and for their classification. Soil interpretations (which judge the suitability of soils for various uses) are also based greatly on morphological descriptions. Because of extensive use of this information, soil scientists are careful to describe soils accurately.

# 9

# SOIL CLASSIFICATION

Soils are as variable as leaves. All the leaves of one tree or of trees of the same kind are similar but no two of them are just alike. From species to species leaves are very different, although they all have the same general characteristics that make them appear as leaves. In the same way, soils of the same kind (series) are similar, but soils of other series are quite different. In some ways, however, the comparison is not valid. Soils grade from one series to another without any sharp boundaries between them, which makes their identification difficult, but it is easy to tell a white oak leaf from a red oak leaf.

The main use of soil classification is to support soil surveys (see Chapter 10). It is impossible for one person who uses a soil to know everything about its use. He or she must depend on the experience of other users and on research; the experience and research from one soil will apply to the same soil in different locations. Soil classification tells us which soils are the same, and soil surveys tell us where they are located.

## Soil Taxonomy

The soil classification system used in the United States is *Soil Taxonomy*. This system is designed to include all soils of the world, and therefore, it is also used in other countries. Soil Taxonomy has six *categories*, or levels within the system. From highest to lowest they are order, suborder, great group, subgroup, family and series. There are eleven orders in the world and more than seventeen thousand series in the United States.

Soil Taxonomy defines several diagnostic horizons and diagnostic materials. Most *soil orders* are defined on the basis of these horizons and materials. *Suborder* definitions are based mostly on soil moisture regimes -- that is, how wet the soil is throughout the year. *Great group* and *subgroup* definitions are based on the presence or absence of certain kinds of soil horizons and on other properties. *Family* definitions are based mainly on particle-size distribution (texture), mineralogy and temperature of subsoil horizons. *Soil series* definitions are based on a wide set of properties, and are named for the local communities in which the soil was first identified. In most published soil surveys the units used for mapping the soils are named for soil series, so this category is the one most familiar to those who use soil information.

## Classifying a Soil

To classify a soil, one needs a morphological description and often laboratory data. The steps to follow in classifying a soil are:
1. Identify the diagnostic soil horizons or diagnostic materials.
2. Determine the soil order.
3. Identify the soil moisture regime.
4. Follow the *Keys to Soil Taxonomy* to key out the lower categories.

### Diagnostic Horizons and Materials

Certain soil horizons, called *diagnostic horizons*, are the result of distinctive soil formation processes. For example, the *argillic horizon* is a clay-enriched horizon that formed by movement of clay from upper to lower horizons. The *mollic* horizon forms by accumulation of organic matter near the soil surface. Other horizons consist of unique kinds of *materials*, such as organic materials, volcanic ash deposits, and high shrink-swell materials. The most important diagnostic horizons and diagnostic materials are described in table 9-1.

Table 9-1.  Major diagnostic horizons and materials in Soil Taxonomy.

| Diagnostic horizons or diagnostic materials | Horizon designation | Properties |
| --- | --- | --- |
| *Surface horizons* | | |
| Mollic epipedon | A, Ap | Dark-colored surface horizon > 10 inches thick |
| Ochric epipedon | A, Ap | Light-colored surface horizon, or thin, dark-colored surface horizon |

Table 9-1 *Continued*

*Subsurface horizons and materials*

| | | |
|---|---|---|
| Andic materials | Bw | Low bulk density, high-activity clay, high P fixation |
| Argillic horizon | Bt | Accumulation of silicate clay |
| Calcic horizon | Bk | Accumulation of calcium carbonate; cemented horizon is a petrocalcic horizon (caliche) |
| Cambic horizon | Bw | Weakly developed horizon |
| Duripan | Bqm | Pan that is strongly cemented by silica |
| Fragipan | Bx | Weakly cemented, brittle horizon with coarse prismatic structure |
| Natric horizon | Btn | An argillic horizon that is high in sodium |
| Oxic horizon | Bo | Highly weathered, rich in Al and Fe oxides, little organic matter |
| Organic material | Oi, Oe Oa | More than 50% organic matter by volume |
| Shrink-swell material | Bss | Material high in swelling clays that forms deep cracks when dry |
| Spodic horizon | Bs | Accumulation of Fe, Al and organic matter |

There is a distinction between horizon designations (tables 8-1 and 8-2) and diagnostic horizons. Horizon designations (for example Bt) represent *some* evidence of a distinctive soil forming process; diagnostic horizons, on the other hand, must show *strong* evidence of that process. Thus, all argillic horizons are Bt horizons, but not all Bt horizons are argillic horizons. Table 9-1 also shows the relationship between diagnostic horizons and horizon designations.

## Soil Orders

The eleven orders of Soil Taxonomy are defined largely on the basis of having certain kinds of diagnostic horizons or diagnostic materials as listed in table 9-2. Soil orders occur in broad zones (fig. 9-1) that depend largely on the climate and organisms (vegetation) soil formation factors, with the exceptions of the Andisol, Histosol, and Vertisol orders, which are more related to distinctive parent materials.

Table 9-2.   Orders of Soil Taxonomy.

| Order | Brief description | Diagnostic horizon or material |
|---|---|---|
| Alfisols | Soils with a subsoil accumulation of silicate clay that are moderately weathered (have a high base saturation) | Argillic horizon |
| Andisols | Soils formed from volcanic materials | Andic material |
| Aridisols | Soils of arid environments | Natric, petrocalcic horizon |
| Entisols | Very weakly developed soils, including many sandy soils | (None) |
| Histosols | Soils formed from organic materials | Organic material |
| Inceptisols | Weakly developed soils, excluding sandy soils | Cambic horizon |
| Mollisols | Soils with thick, dark surface horizons that are high in organic matter content | Mollic horizon |
| Oxisols | Very highly weathered soils of tropical areas that are high in iron- and aluminum-oxide minerals | Oxic horizon |

Table 9-2 *Continued*

| Spodosols | Soils with a subsoil accumulation of aluminum, organic matter and usually iron | Spodic horizon |
|---|---|---|
| Ultisols | Soils with a subsoil accumulation of clay that are highly weathered (have a low base saturation) | Argillic horizon |
| Vertisols | Soils that undergo much shrinking and swelling | Shrink-swell material |

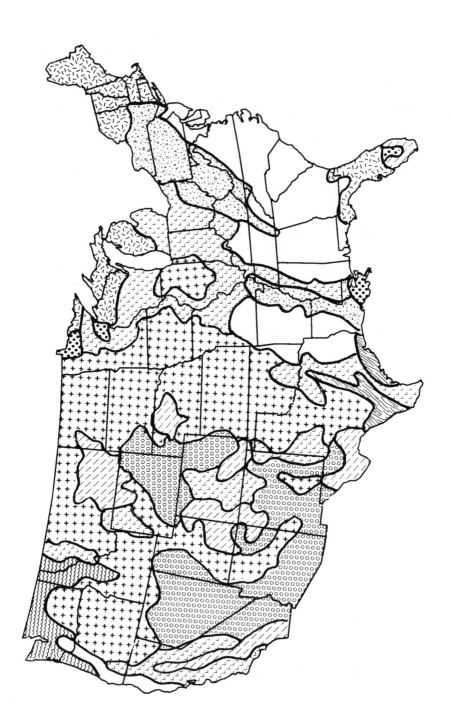

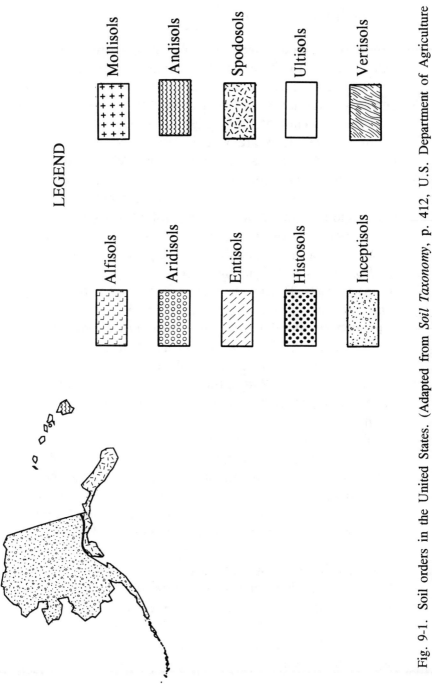

Fig. 9-1. Soil orders in the United States. (Adapted from *Soil Taxonomy*, p. 412, U.S. Department of Agriculture Handbook no. 436, 1975, Washington, DC: U.S. Government Printing Office.)

LEGEND

Alfisols
Aridisols
Entisols
Histosols
Inceptisols

Mollisols
Andisols
Spodosols
Ultisols
Vertisols

Spodosols, Inceptisols, Alfisols, and Ultisols are the extensive soil orders in the eastern part of the United States (fig. 9-1). Spodosols are found mainly on the sandy deposits in the northern Great Lakes States, New England and Florida. They formed under forest vegetation, mostly conifers. Inceptisols are weakly developed soils that are mainly on the steeper slopes of the Appalachian Mountains and on the flood plains of the larger rivers, especially the Mississippi. Alfisols and Ultisols formed in humid climates where the downward percolating rainfall translocated clay from upper to lower horizons. Alfisols formed on the younger parent materials that were deposited by glaciers in the northern states. They are also in a narrow belt east of the Mississippi River where they formed on silty material that was washed out of glaciers, carried down the river, deposited on flood plains, and then blown onto upland areas as loess. Ultisols formed on the older, nonglaciated landscapes and are more leached than Alfisols because they formed for a much longer time and under higher temperatures, in which chemical weathering reactions are more rapid.

Mollisols, Entisols, Alfisols and Aridisols are the extensive soil orders of the western United States. Mollisols are found on the grasslands, mainly in the Great Plains and the Columbian Plateau. They also extend eastward through Iowa, Illinois and eastern Indiana in an eastern extension of the major prairie called the Prairie Peninsula. Aridisols are in the very dry parts of the Great Plains and the range-and-basin areas. Entisols are soils with very little development. They formed where the soil is very steep as in the mountains, very sandy as in the Sand Hills, or very young as in the California trough. Much of Alaska has soils with permafrost, permanently frozen soils, that are classified as Inceptisols.

The Histosols, Vertisols and Andisols are less extensive because they formed on unique parent materials. Histosols are organic soils that form in shallow lakes. They are in the northern Lake Sates, Florida and the Mississippi Delta. Vertisols formed on certain deposits that weather to form soil minerals that undergo much shrinking and swelling. They are most extensive in Texas. Andisols formed on volcanic deposits in the Northwest.

### Soil Moisture Regimes

The next major step in classifying a soil is to determine its soil moisture regime. Four of the regimes depend mainly on the regional climate (fig. 9-2):

Udic -- soils of humid climates

Ustic -- soils of semi-arid climates

Xeric -- soil of Mediterranean climates

Aridic -- soils of arid climates (deserts)

In the udic and ustic soil moisture regimes, the rain is uniform throughout the year or it comes mainly in the summer. The xeric regime is in Mediterranean climates where most of the rain falls during the winter and the summers are dry. In the aridic regime there is very little rain during any season.

The aquic soil moisture regime depends on local factors, such as parent material and topography, that are responsible for maintaining high water tables in soils. Soils with aquic moisture regimes receive water that runs off higher areas of the landscape and/or have horizons with slow hydraulic conductivity. They are scattered among soils with other moisture regimes but are more common in udic areas than in ustic and aridic areas.

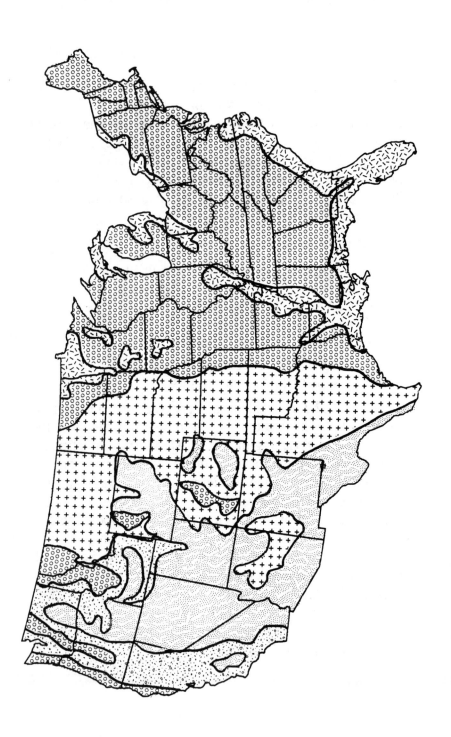

# LEGEND

Moisture regimes controlled mainly by regional climate:

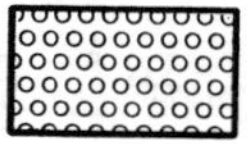

 Udic: soils of humid climates

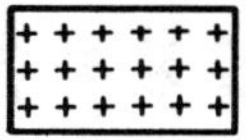

Ustic: soils of semi-arid climates

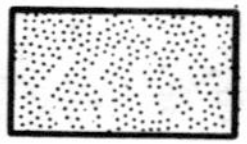

Aridic: soils of arid climates

 Xeric: soils of Mediterranean climates

Moisture regime controlled mainly by local conditions (slow permeability, low landscape position):

 Aquic: soils that are periodically saturated and reduced

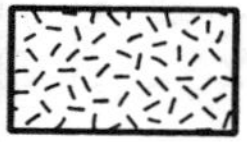

Fig. 9-2. Soil moisture regimes in the conterminous United States. (Adapted from preliminary maps produced by the Soil Conservation Service, U.S. Department of Agriculture.)

A soil is placed in the lower taxonomic categories by following the soil keys, similar to the way one keys out a plant using botanical keys. Soil series, the lowest category, are defined by a representative pedon description and statements about the allowable range in soil characteristics and how that series related to others in the landscape.

## Nomenclature

Soil Taxonomy uses nomenclature based largely on Greek and Latin formative elements. Many of these formative elements are also used in other words common in the English language. For example, names for wet soils contain the formative element *aqu*, as in aquarium. Soils with high base saturation, or high fertility, may contain the root *eutr*, which also appears in the word eutrophication which describes the condition of lakes when they become enriched with plant nutrients. The nomenclature system of Soil Taxonomy is easy to learn and apply. Here are the system's main rules:

1. The eleven orders (table 9-2) all have names ending in *sol* (e.g., Alfisol, Mollisol).
2. A suborder name has two syllables, the last being the formative element of the order to which it belongs. For example, Aqualfs are wet Alfisols, and Udolls are Mollisols of humid climates.
3. A great group name consists of a prefix added to a suborder name (e.g., Ochraqualfs, Argiudolls).
4. A subgroup name is formed by adding an appropriate adjective, as a separate word, to a great group name (e.g., Aeric Ochraqualfs, Typic Argiudolls).
5. A family name is designated by modifiers that describe particle-size distribution, mineralogy and temperature (e.g., fine-silty, mixed, mesic, respectively).

6. A series is named for the locality in which a particular soil was first described, for example, the Cincinnati series.

Table 9-3 lists some of the formative elements used in names of suborders, great groups and subgroups. By knowing the meanings of these formative elements and the general descriptions of the soil orders in table 9-2, one can deduce the important properties of a soil series from its classification.

Table 9-3.  Formative elements used in Soil Taxonomy

| Element | Derivation* | Common English word | Meaning |
|---|---|---|---|
| aeric | (G) *aerios*, air | aerial | Freely drained (not waterlogged) |
| aqu | (L) *aqua*, water | aquarium | Reduction due to waterlogging |
| bor | (G) *boreas*, north | boreal | Cool |
| camb | (L) *cambiare*, to exchange | change | A cambic horizon (weakly developed B horizon) |
| dys, dystr | (G) *dys*, ill, infertile | dystrophic | Low base saturation |
| eu, eutr | (G) *eu*, good; *eutrophos*, fertile | eutrophic | High base saturation |
| fluv | (L) *fluvius*, river | fluvial | Flood plains |

*G = Greek derivation;  L = Latin derivation;  E = English word

Table 9-3 *Continued*

| frag | (Modified from L) *fragillis*, brittle | fragile | Presence of fragipan |
|------|------|------|------|
| hapl | (G) *haploius*, simple | haploid | Minimum set of horizons |
| lithic | (G) *lithos*, stone | lithosphere | Shallow to bedrock |
| ochr | (G) *ochros*, pale | ocher | Ochric epipedon (a light-colored surface horizon) |
| orth | (G) *orthos*, true | orthoscopic | The common ones |
| psamm | (G) *psammon*, sand | psammite | Sand textures |
| typic | (E) typical | typical | Typical of the great group |
| ud | (L) *udus*, humid | udometer | Udic (humid) soil moisture regime |
| ust | (L) *ustus*, burnt | combustion | Ustic (semi-arid) soil moisture regime |
| xer | (G) *xeros*, dry | xerophyte | Xeric (Mediterranean) soil moisture regime |

For example, here is the information about the Avonburg soil that you can infer from its soil family name, Aeric Fragiaqualf, fine-silty, mixed, mesic:

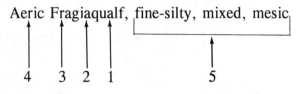

1. Order: Alfisol (-**alf** is the formative element of the Alfisol order) -- the soil has a B horizon of clay accumulation with a fairly high base saturation.
2. Suborder: Aqualf (-**aqu**-; *aqua*, water) -- the soil has evidence of water logging.
3. Great group: Fragiaqualf (**Fragi**-; *fragilis*, brittle) -- the soil has a fragipan, a subsurface layer that is brittle when moist and hard when dry.
4. Subgroup: Aeric Fragiaqualf (**Aeric**; *aerio*, air) -- the soil is less wet and, consequently, better aerated than most of the soils of the great group.
5. Family modifiers: **fine-silty** -- subsurface horizons contain between 18 and 35% clay and less than 15% coarser than very fine sand; **mixed** -- no single mineral predominates in the whole soil; **mesic** -- the average annual soil temperature corresponds to that of the Corn Belt (see below).

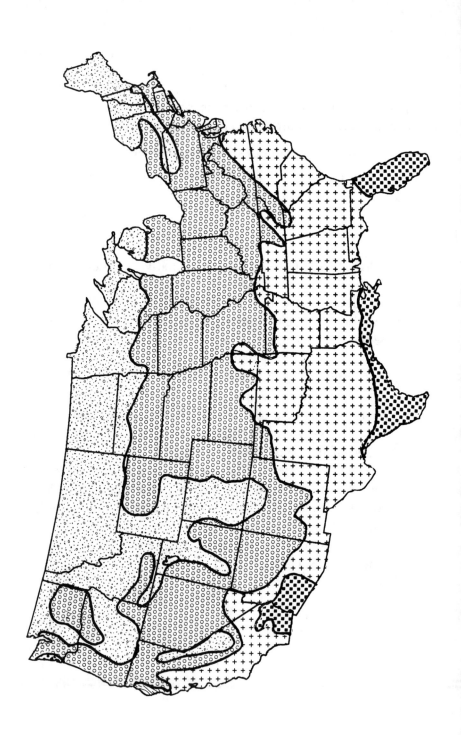

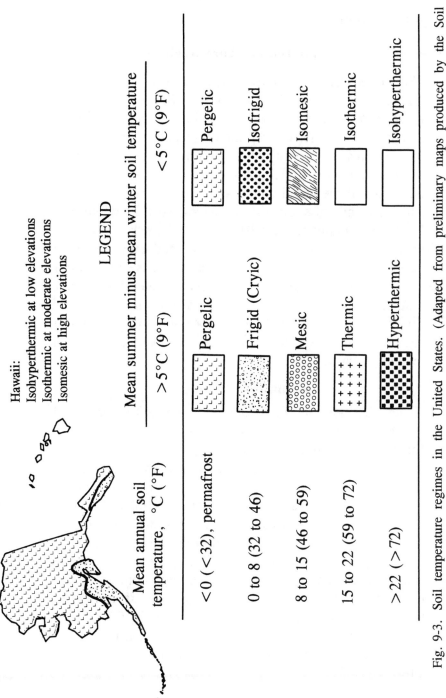

Fig. 9-3. Soil temperature regimes in the United States. (Adapted from preliminary maps produced by the Soil Conservation Service, U.S. Department of Agriculture.)

## Soil Temperature Regimes

Soil temperature regimes depend mainly on latitude and altitude (fig. 9-3). In the United States, temperature regimes range from hyperthermic in the south to pergelic in Alaska. The thermic regime corresponds approximately to the climatic zone in which cotton is grown. The mesic regime comprises much of the Corn Belt. The southern boundary of the frigid temperature zone is near the natural southern limit of some forest species, such as white pine. In Alaska, many soil horizons are frozen throughout the year, a condition called permafrost, and they are in the pergelic temperature regime. In all of these temperature regimes there is a difference between winter and summer temperatures. In some coastal areas, however, this difference is very small, and hence the soil temperature regimes have *iso* (equal) prefixes, as documented in the legend of figure 9-3.

## Using Soil Taxonomy

There are essentially two ways to use Soil Taxonomy: (1) to classify a soil and (2) to deduce the properties of a soil for which the classification is known. More people will use Soil Taxonomy in the second way. For example, assume that someone is trying to sell you a house lot and you determine from a soil survey that it is on a Fluventic Eutrochrept soil. Knowing a few of the terms you can deduce major soil properties. From the information in this chapter you deduce that the soil has weakly developed subsoil horizons (Inceptisol) and a light-colored surface horizon (**ochr**ic epipedon). It has high base saturation and high fertility (**eutr**) and is on a flood plain (**Fluventic**). This soil might be a good one for a garden, where occasional flooding could be tolerated, but not for a home site.

# 10

## SOIL SURVEYS

The best way to apply the principles of soil science to managing specific tracts of land is to use soil surveys. They provide a great deal of information about our land resources. In fact, for many areas they contain more natural resource information than any other source. Soil surveys are published for individual counties or similar geographic areas. They contain five kinds of information: a general description of the area, technical information about how the soils formed and how they are classified, detailed soil maps, descriptions of soils and landscapes, and soil interpretations -- suggestions about the suitability of soils for various uses and how they are expected to respond to specific practices. It is the goal of this chapter to introduce you to soil surveys so that when the opportunity arises you can use them to plan community resources, conserve natural resources, or decide how to use land that you own or manage, whether it be a farm, ranch or homesite.

### General Information

The main feature of this section of a soil survey is a general soil map of the survey area. It also includes information about the geology, land forms, and climate of the area, and it may include sections with cultural information, such as population trends, major industries, and transportation. The general map reflects the major kinds of land forms and soil parent materials. This map and the written descriptions of the map units are used for general planning, such as zoning, developing resource inventories, conservation planning, and deciding where to target certain activities, such as erosion control or sales of certain farm equipment or chemicals that are especially suited to the soils of that area.

## Detailed Soil Maps

Detailed maps are the heart of a published soil survey. These maps are made on air photos, mainly at a scale of about 1:15,000 to 1:30,000 (four inches for every one mile to two inches for every one mile). A soil scientist usually begins the mapping process by examining overlapping air photos with a stereoscope to see the landscape in three dimensions and to get an overview of the terrain. He or she then develops a soil map by walking over the area, observing the vegetation, landforms, geology and other natural features, and boring holes with a soil auger to study the different kinds of soil horizons. From the auger sample, the soil scientist observes the kinds of horizons and their textures, colors, reactions and other attributes. Using this background, he or she decides what area on the ground is homogenous in kind of soil, slope, and amount of erosion, and draws a line on the air photo to represent the area. The soil scientist may revise this line by again using a stereoscope. The area enclosed by a line on a map is called a map *delineation,* and, collectively, similar delineations constitute a *map unit,* which is identified by a *map unit symbol* such as those listed in table 10-1.

Table 10-1.   Example of part of a legend from a soil survey.

| Symbol | Name |
| --- | --- |
| AvA | Avonburg silt loam, 0-2% slopes |
| AvB2 | Avonburg silt loam, 2-4% slopes, eroded |
| BeD3 | Beasley-Rock outcrop complex, 12-25% slopes, severely eroded |
| BnC2 | Bonnell silt loam, 6-12% slopes, eroded |
| BnC3 | Bonnell silt loam, 6-12% slopes, severely eroded |
| BnD2 | Bonnell silt loam, 12-18% slopes, eroded |
| BnD3 | Bonnell silt loam, 12-18% slopes, severely eroded |
| BnE | Bonnell silt loam, 18-45% slopes |
| Co | Cobbsfork silt loam |
| Ho | Holton loam, occasionally flooded |

From *Soil Survey of Jefferson County, Indiana* by A. K. Nickell, 1985, Washington, D.C.: U.S. Government Printing Office.

Soil map unit names contain three kinds of information: the kind of soil (usually the soil series), the steepness of slope, and the amount of past erosion. For example, in the map unit symbol:

AvB2

Avonburg silt loam (Av) ⅃↑ ↑ ↑
2 to 4% (B) slopes ⅃
Moderate (2) erosion ⅃

The first capital letter and the lower case one that follows (Av) represent the map unit name and the texture of the surface horizon -- in this case, they refer to the Avonburg soil series. The second capital letter (B) represents the slope class 2-4%. The slopes of the map units listed in table 10-1 range from 0-2% (A) to 18-25% (E). Symbols without a slope letter, such as Co, are nearly level soils. A final number (2) indicates the degree of erosion. Three degrees of erosion are recognized -- slight (no number), moderate (2), and severe (3).

Table 10-1 shows that there are five map units named for the Bonnell series, two with 6-12% slopes in which one is moderately eroded and one is severely eroded, two with 12-18% slopes that also differ in erosion, and one with 18-45% slopes that is slightly eroded. One may question why the steepest soil has the least erosion. It is because the soils with slopes less than 18% were mainly cleared of trees and farmed, and those with slopes steeper than 18% were always in trees and thus not eroded. The name of the map unit suggests that soils represented by the symbol Ho are occasionally flooded.

The soil map also shows water drainageways and other special features such as stones on the surface. Within map delineations there are always areas that differ in kind of soil, slope or erosion from the dominant condition but are too small to be mapped out. These anomalous areas are called *inclusions*. During the course of mapping, soil scientists collect many notes on pedons. They summarize this information for a particular kind of soil and then select one site that best represents the soil about which to write a detailed morphological description.

## Technical Information

The technical information section is written mainly for those with some scientific background in soil science, such as provided in this book. It tells how properties of the soils depend on the five factors of soil formation and explains how the soils formed. This section of the survey includes detailed morphological descriptions from representative sites. These pedon descriptions include information about the color, texture, structure, reaction, and so forth, for each horizon. They also explain the variability that is expected within each kind of soil, and the relation of each soil to others in the survey area. For most soil surveys, these "kinds of soils," as demonstrated in the example above, are soil series. In some surveys, however, soil mapping might be done by "kinds of soil" at other categorical levels in Soil Taxonomy, such as soil families. Soil families are often designated when mapping areas where the soils are not used intensively, such as in grazing land or forest land. When mapping by soil series, a soil scientist will also list the classification of each soil at the family level in the technical information part of the soil survey. Furthermore, this section includes tables of data on chemical, physical and engineering soil properties.

## Description of a Soil Map Unit

After much of the mapping is completed, the soil scientist prepares a narrative map unit description. In contrast to a description of a soil at a certain *point*, the map unit description represents an *area* of soil. Typically, the map unit description presents a condensed analysis of a representative pedon in nontechnical terms, a summary of the soil landscape, the amount and kinds of inclusions, and a summary of the hydrological properties of the soil, such as permeability and water-holding capacity. It also includes statements about the suitability of that kind of soil for various uses.

## Soil Interpretations

Another major part of the soil survey describes the suitability of the soils for various uses and predicts how the soils will respond to certain kinds of management. These are called *soil interpretations*. Some interpretations are given as numerical ratings, such as the predicted yields for certain crops or the growth rate of trees. In other cases, suitability is described in terms of the *limitations* of the soil for a particular use. A *slight* limitation means that the soil is generally suitable for that use and that restrictions are generally minor and easy to overcome. A *moderate* limitation means that the soil is not favorable for the use, but with special design, planning and financial investment the soil could be made suitable. A *severe* limitation generally means that the restrictions are so severe that they cannot be overcome, or that the cost of overcoming them is so high that it is not practical to do so.

Table 10-2 lists some examples of soil interpretations found in a soil survey of Jefferson County, Indiana. Only about 20% of the kinds of interpretations given in that publication are listed in the table. In other areas of the country, different kinds of interpretations are given, for example, cranberry yield in Wisconsin, cotton yield in Texas, and growth rate of Douglas fir trees in Washington.

Table 10-2.  Examples of interpretations for the soil map unit AvB2, Avonburg silt loam, 2-4% slopes, moderately eroded.

| Soil use | Rating | Comments |
|----------|--------|----------|
| Land capability class | IIe | Suitability for crop production on a scale of I (best) to VIII (poorest); e = soil has erosion hazard |
| Expected corn yield | 100 bushels/acre | Using good management |
| Pasture productivity | 7.2 AUM (animal-unit-months) | One AUM is the amount of forage needed to feed one cow (or one horse, or five sheep, etc.) for one month |
| Woodland productivity | 3d | Ordination symbol based on a scale of 1 (most) to 5 (least) productive; d = soil has restricted rooting depth |
| Seedling mortality | Moderate | For tree planting |
| White oak growth | Site index = 70 | Expected height in feet of 50-year-old trees |
| Trees to plant | Red maple | [Six other species suitable to the soil are also listed] |
| Camp area | Severe | Too wet for campsites |
| Wildlife habitat | Good | Based on wild herbaceous plants |

Table 10-2  *Continued*

| | | |
|---|---|---|
| Dwellings without basements | Severe | Because of wetness |
| Septic tank waste absorption field | Severe | Because of wetness and low permeability of subsoil |
| Source of sand | Improbable | Poor chance of finding sand |
| Pond reservoir | Moderate | Pond might leak because of seepage |
| Water table depth | 1 to 3 feet | During wettest season |

From *Soil Survey of Jefferson County, Indiana* by A. K. Nickell, 1985, Washington, D.C.: U.S. Government Printing Office.

# 11

# SOIL EROSION AND ITS CONTROL

Soil erosion is the process of detaching and removing soil materials from their original sites. Wind and moving water supply the main forces responsible for this phenomenon. In humid areas, water is the major cause of erosion; in semi-arid areas, both water and wind are responsible for much erosion damage.

## Soil Erosion by Water

Water erosion is made up of two components: *detachment* and *transport*. Detachment of soil particles from the ground is brought about mainly by moving water -- rain and surface runoff -- although several other factors may also be active, such as freezing and thawing, wetting and drying, and mechanical action of tillage implements or vehicles. In all cases energy has to be expended to detach the soil particles.

Rain drops hit the ground with velocities between 10 and 20 miles per hour, while surface runoff -- outside of rills and gullies -- usually travels not more than 1/2 mile per hour. The impact of rain drops, especially large, fast-traveling rain drops, on the unprotected soil causes much more detachment of soil than the surface water moving across the average field. However, wherever much water concentrates in draws or gullies, the velocity of runoff increases and serious detachment results.

After soil particles have been detached, they can be transported away. Low flow velocities will carry off large amounts of soil, as small soil particles become suspended in runoff water. The smaller the soil particles, the slower they settle and the more readily they are washed away. This trend is at times countered by the tendency of some soil particles to cling together to form aggregates. Clay particles often do this. Silt particles mainly act individually. For this reason, soils high in silt are most erodible. Organic substances are also easily transported by water because of their low density.

Surface runoff is not the only means of soil transportation. Rain splash may kick up much soil and transport it with the wind or down the slope.

### Forms of Erosion by Water

The gradual removal of thin layers of soil from an area is called *sheet erosion. Rill erosion* is the washing away of soil in shallow channels that do not penetrate the normal plow depth. The channels can therefore be smoothed out completely by cultivation. For this reason, the effect of rill erosion is essentially the same as that of sheet erosion. The formation of deep gashes in the ground at places where runoff concentrates is *gully erosion*. The sliding down of whole masses of soil over water-saturated and lubricated strata is called *landslides*.

Of these, sheet and rill erosion are by far the most important. Soil losses by sheet and rill erosion outweigh the other two forms probably one hundred to one for the country as a whole, although the spectacular nature of gullies creates more public concern. As a matter of fact, the formation of gullies indicates that erosion has advanced so far that the use capability of the land is reduced, while the insidious nature of sheet erosion causes many farmers to overlook the danger until it is too late. Mass movement of soil is only of local importance, occurring mainly in areas of high rainfall where level and slightly inclined geologic strata of different permeabilities alternate.

### Factors Affecting Soil Erosion by Water

The rate of erosion is affected by four kinds of factors: climate, soil properties, slope, and surface cover and land use.

*Climate.* The most important climatic factor is the rainfall. Hard-beating rains with large drops cause serious erosion hazards. The total annual precipitation is of little significance in this respect. Hard rain detaches much soil (fig. 11-1) and is not absorbed as fast as it falls, and hence, runs off, transporting soil off the field. Gentle rain, even if of long duration, causes little erosion. Wind may drive raindrops against the soil surface at high velocities; otherwise, wind is of only indirect importance in erosion by water. High temperatures tend to dry out surface soils and to make them readily detachable. In arid areas plant growth is so scant that erosion is an ever-present menace, even without the interference of people. Also, freezing and thawing detaches soil particles and exposes them to removal by runoff. As long as the soil is frozen solid, no erosion occurs. This is one reason why there is less erosion in the northern states than in the southern states. The climate of an area dictates to a certain extent the type of land use and therefore affects the potential erosion.

Fig. 11-1. Rain drop falling on unprotected soil.

*Soil Properties.* As stated above, the two main processes of soil erosion are detachment of soil particles and their transport. Soils that resist these two processes are most resistant to erosion. To resist detachment, the soil must be made up of water-stable aggregates, that is, the ultimate soil particles must be grouped into masses that cling together even when they are submerged in water. A fair content of clay acts as the binder for coarser soil particles, but it takes the glue provided by the microbes as they decompose organic residues to keep these aggregates together under the impact of water. All the factors that make for production of much organic matter and for its ready decomposition help to make the soil resistant to detachment. Also, to be resistant to transportation, soil particles or aggregates must be so large that they cannot be floated off easily. The larger a soil particle, the faster it sinks in water. Soil aggregates of sand size do not stay in suspension long. Silt, on the other hand, may be carried far before it is settled out. Nonaggregated clay remains in suspension almost indefinitely. Another factor affecting the transportation hazard is the infiltration capacity of the soil, that is, its ability to absorb water. Where more water enters the soil, less runoff is available to transport soil particles.

*Slope.* *Steepness, length* and *shape* of the slope affect the rates of runoff and erosion. Increasing the steepness of the slope greatly increases the velocity of runoff and hence the rate of erosion. Also, the transportation by splash erosion increases with the steepness of the slope. With increasing length of slope, the amount of runoff water accumulates downslope to create more erosion hazard. This is especially true on slightly pervious soils. Long slopes cause water to collect in draws and to cut gullies. More sheet erosion occurs on convex (bulging) slopes than on concave (depressed) slopes, because the convex part of the slope is usually drier and lower in organic matter and less vegetated. Another reason for this is that on the concave slopes, the steepness decreases. This slows down the moving water and some of the suspended load is deposited on the lower part of the slope. On convex slopes, the steepness increases downslope and the velocity and carrying capacity of runoff flow increase toward the lower end of the slope.

*Surface Cover and Land Use.* A dense cover of vegetation with plant residue on the ground, as found under undisturbed natural conditions in forest or prairie land, provides complete protection from erosion (fig. 11-2). Only where the soil is exposed along creeks, by uprooted trees, through landslides, or through any other natural cause can erosion occur in forest or prairie land -- that is, until humans upset the natural equilibrium. Clean tillage systems leave the soil unprotected for considerable periods of time. Tillage also causes more rapid decomposition of organic matter. Therefore, clean tillage results in accelerated erosion.

Fig. 11-2. Rain drop falling on soil protected by crop residue.

The actual effect of agriculture on soil erosion depends largely on the amount of surface cover and the intensity of tillage. It is quite obvious that a cultivated field without any cover of plants or plant residues represents a much greater erosion hazard than a dense bluegrass pasture or undisturbed woodland. Many intermediate conditions of soil cover exist. These include, in order of increasing protection for the soil:

Fallow
Row crops
Large seeded legumes

Small grains
Small seeded legumes
Grass-legume meadows
Permanent vegetation

The erosion hazard created by a certain land use depends on the plants and the plant residues covering the ground, and on the management of the plants and soil that go together with the particular land use. Such items as the rate of growth (as affected by soil fertility and climate) the period the crops are on the ground, and the disposal of the crop residues determine to quite an extent the erodibility of the land use. The sequence of clearing the land, clean tillage up and down the slope, gully formation, and soil conservation farming is illustrated in figures 11-3 to 11-6.

Fig. 11-3. Clearing the land.

Fig. 11-4. Clean tillage up and down a slope.

Fig. 11-5. Gully formation--exhausted land.

Fig. 11-6. Soil conservation farming.

The rate at which the fields of the United States erode, as accelerated by farming, is by and large much faster than the formation of new soil. It has been estimated that soil loss from moderately sloping cropland in the United States is around 4 to 6 tons per acre per year as an average for all crops grown. Losses from corn and other open-tilled crops are from 2 to 100 tons per acre per year; from small grain they are from 2 to 16 tons per acre per year; and from well-established meadows and well-managed pastures, the losses are a small fraction of a ton per acre per year.

The total amount of erosion in the United States is staggering. Over 3 billion tons of soil are lost annually. This corresponds to over 3 million acres of top soil (at 1000 tons of soil/acre-7 inches), an area about the size of Connecticut. Additionally, the average depth of top soil has decreased in the United States from 8 to 5 inches. Despite the statistics, we actually are making progress in fighting soil erosion. In the 1966 edition of this book, the estimate of soil erosion was 5 billion tons per year. We have a long way to go, however. As mentioned above, in most crop-land, soil is still eroding faster than it is being formed.

## *Sedimentation*

Erosion is an intermittent process. Soil that is carried from its original location in the field in one storm fre-quently is deposited not far away, only to be picked up in another storm and carried on farther down toward the sea. The finer the soil particles, the farther they are carried each time. Only clay, dispersed into its ultimate particles, can stay in suspension indefinitely -- until the salt content of the sea or some other similar agent coagulates the clay particles and causes them to settle out.

Sedimentation of eroded material may be beneficial or detrimental, depending on the nature of the sediment and on the place of deposition. Sediment deposited on a flood plain soil, for example, adds to its depth and its fertility. In most cases, however, sedimentation is detrimental, such as when it covers roads and fills reservoirs.

## Soil Erosion by Wind

Wind also causes much soil erosion. When the wind blows across a bare, dry soil, it picks up some particles. The particles are accelerated into the air stream. Some also fall toward the surface, where they strike other particles and aggregates with great force and dislodge more particles into the wind. These are also accelerated (fig. 11-7). Thus, once soil particles start to move, the process is self-perpet-uating. Fine-sand and silty soils low in organic matter are most subject to wind erosion. Coarser sand particles are too heavy to be picked up easily, and finer particles tend to cling together in aggregates, especially if moisture is present. Wind erosion is especially severe during times of drought, when the soil lacks moisture and the vegetative cover is sparse. One of the most spectacular examples of wind erosion occurred in the 1930s in the Great Plains, which became known as the Dust Bowl. Farming had spread to that area during previous wetter periods, and vast areas were plowed up. When the drought hit, vegetation withered, and the soils blew. They covered fences and houses and choked animals and people. Huge clouds of dust were carried all across the continent, to the extent that they blotted out the sun in New York City and Washington D.C.

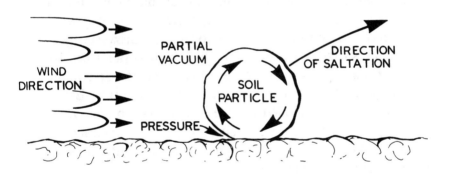

Fig. 11-7. Effect of wind on movement of soil particles.

## Soil Conservation

Practically everywhere, topsoil grows better crops than subsoil. There are several reasons for this. Plant residues accumulate on top of the soil and, upon decay, enrich the surface soil. In the course of soil development clay usually washes down the profile and organic matter accumulates in the surface layer (A horizon), bringing about a mellow structure that is readily permeated by plant roots, water and air. Much of this clay is deposited in the subsoil (B horizon), making the subsoil more dense and more resistant to root penetration. Roots do penetrate through the voids around structural aggregates and have access to the moisture stored in the subsoil, but this moisture is not as readily available to plants as the topsoil moisture. For ages, plant roots have foraged throughout the soil. This combination of mellow, fertile soil on top and heavier, water-retentive soil below -- as created by nature -- is the best medium for plant growth we can imagine. Agriculture can remove the A horizon and expose the B horizon. The B horizon, as stated above, is low in organic matter and plant nutrients and is frequently high in clay and slow in absorbing water. Much of the rain falling directly on soil B horizons runs off over the surface. The result is that crops suffer from lack of nutrients and lack of water. It is a matter of common observation that yields on eroded soils are inferior. Experiments in the Middle West have shown that on the average corn yields decrease by 5 bushels per acre for each inch the surface soil is less than 12 inches thick. For these reasons, we must save our soil resources.

### Aims of Soil Conservation

Every farmer must aim to conserve the surface soil to maintain the productive capacity of his or her land. It is impossible to cultivate land without any loss of soil, so permissible rates of erosion have been established by soil conservation agencies. A *soil loss tolerance*, or *T value*, is said to be the amount of soil loss that can be tolerated each year while still maintaining high soil productivity. For most cropland, T values are around 4 to 5 tons per acre. An acre of soil 7 inches thick weighs about 1000 tons. An annual loss of 4 tons is equivalent to the total removal of the upper 7 inches in 250 years. This rate of soil loss probably is much greater than the rate that soil forms from its parent material. Since many soils are eroding at rates much greater than 5 tons per acre, however, the immediate goal should be to bring erosion down to T values. The long-range goal, on the other hand, should be to determine rates of soil formation and to develop and use soil conserving methods that will limit erosion rates to soil formation rates. Such techniques most likely will require that soils have some kind of cover at all times.

## Reducing Detachment

Erosion can be minimized by limiting both the detachment and transport processes. Rain drops travel at high velocities as they strike the ground and so have a great deal of energy. The first defense against detachment is to reduce the force of rain drop impact on the soil by keeping the soil covered at all times with living plants or plant residues which absorb most of this energy. The next task is to encourage the formation of soil aggregates that will not break into individual particles when hit by raindrops. These water-stable aggregates are held together mainly with organic materials, so additions of organic matter to the soil is encouraged. The organic matter that provides surface cover also serves this purpose.

In farming operations, soil cover is best maintained by growing perennial crops such as grasses and legumes. When clean-tilled crops such as corn and soybeans are grown, cover is maintained by leaving the crop residue on the surface. If this cover is insufficient to control erosion, a winter cover crop should be planted in the fall after harvest. A winter crop will grow enough during the late fall so that its roots will help bind soil particles together and its tops will provide surface cover over winter and early spring when soils are very subject to erosion. A winter crop can be killed with herbicides or turned under before the next crop is planted in the spring. It also adds organic matter to the soil.

In traditional farm operations, the surface soil is first turned over by a moldboard plow or by chisel plows, which bury crop residue. Then the soil is further tilled by several trips with disks, and the new crop is planted into the finely-pulverized soil. This soil is very erodible. In conservation tillage methods, all or most of the crop residue is left on the surface, and the new crop is planted through the residue. This reduces soil pulverization (leaving more aggregates), reduces compaction and saves fuel costs by minimizing the number of tillage operations. Live plants and crop residue also slow down runoff. Weeds, and often winter cover crops, are killed with chemicals in the no-till system.

### Reducing Transport

Transport of eroded soil particles is reduced by decreasing the amount and velocity of runoff water. The amount of runoff can be reduced by allowing more of it to infiltrate into the soil. Infiltration, in turn, is increased by maintaining good soil structure and by creating large pores through which water can flow. Addition of organic matter to the soil decreases detachment and promotes the formation of a loose, porous soil structure that conducts water quickly. Plant roots growing in the soil also create pores through which water can flow. Other pores are made by earthworms, which depend on organic matter for their food.

The velocity of runoff is reduced by creating obstructions to water moving straight downhill -- specifically, by maintaining vegetative cover, by conducting all farming operations on the contour (along the slope, not up and down; Fig. 11-6), by leaving the soil surface rough with many small depressions, and by building structures such as terraces to break up long slopes.

## Controlling Wind Erosion

Controlling wind erosion is similar in many respects to controlling water erosion. Soil cover reduces wind erosion by absorbing the energy of wind-blown particles and by reducing detachment of particles from aggregates. Because it tends to perpetuate itself, wind erosion builds up across long, unobstructed sweeps of land. This cycle can be broken by keeping vegetative cover on the soil at all times and by planting rows of trees in windbreaks.

## Soil Conservation and Society

Humankind continues to increase in numbers -- by 93 million (the equivalent of 12 New York Cities) annually. Today, there are 5.6 billion of us, and as the numbers of tillable acres steadily diminish through erosion and other causes, there is no alternative but to bend every effort to conserve the soil that is left. A large part of the earth's population is either ill-nourished or actually starving. The conservation of the productive capacity of the land becomes the responsibility of every farmer.

Research into better ways of controlling erosion and into more efficient uses of the land and crops brings us steadily closer to successful soil conservation. The economic importance of soil conservation and its value in the nutrition of humankind has received so much attention that in both the city and the country, *conservation farming* has become a synonym for *good farming*.

The United States has destroyed more land faster than any other nation. Fortunately, it has adopted federal and state programs of soil conservation -- which deserve the support of every citizen. Some countries have energetically attacked the soil erosion problem for many years, and others are now following our system of soil conservation organization.

Full success of a soil conservation program can only be realized when this subject has been taught to today's children by teachers who appreciate the problem of soil conservation and when a soil stewardship attitude has been created in the entire generation.

# 12

## SOIL AND THE ENVIRONMENT

Soil is the central component of the environment. It is here that the physical, chemical, biological and climatic factors merge. Earlier chapters showed how soil properties depend on the factors under which the soils developed -- parent material, topography, climate, organisms and time. This chapter explains how soils interact with their present-day environments, how they can become degraded, and how we might care for them to minimize degradation and protect other parts of the environment.

### The Natural Soil Environment

In this section we will first discuss how some general principles discovered through studies of soil formation can guide us in using our soil resources in the future, and then we will examine some aspects of the natural soil environment that relate directly to modern-day environmental concerns.

### *General Principles*

Soil formation times are measured in thousands of years. Most soils formed under natural conditions for more than tens of thousands of years, many for hundreds of thousands of years, and some for millions of years. Given the same natural conditions, it is likely that these soils will last for thousands more years. As a rule, then, *if we wish to use our soils for an unlimited time, we should use them in a manner similar to the way they existed in their natural states*. This is the general principle of soil care.

By studying soil formation, we also learn that during thousands of years of soil formation, soil properties changed very slowly in response to environmental changes. Today, many properties continue to change slowly in response to changes in their present environments, so the damage we do to a soil might not be immediately apparent. Once a soil is damaged, however, it usually takes a long time to heal the scars. This leads to another general principle of soil care: *We should look for early signs of soil damage and adjust management plans to reduce the harm before it becomes so severe that it will take many years to repair it.*

An example of a process that proceeds slowly -- where it is important to look for early signs of damage -- is the movement of certain chemicals, such as nitrate, through the soil to the ground water. Water moves very slowly through the subsoil and lower strata of many soils, and it takes many years for the water that infiltrates the soil to reach the ground water reservoir. If excessively high rates of nitrogen are applied to a soil, some of it will move downward as nitrate with the soil solution. Because the water moves so sluggishly, the nitrate might not appear in the ground water for many years, perhaps for even hundreds of years. Once it does appear, however, it will take a similarly long time to flush out the excess nitrate after high application rates have ceased.

### Surface Cover

In their natural settings, essentially all soils have some kind of surface cover. Mostly it is vegetation, but it might also be a gravely mulch, such as in the desert. This cover protects the soil from erosion by preventing raindrops from directly striking erodible soil material. In this way, soil cover reduces the detachment of soil particles and the transport of detached particles, the two major processes of soil erosion (see chapter 11). The cover also creates obstructions, or small dams, that slow down the water that runs over the surface of the soil, and vegetation provides channels through which water can enter the soil.

Originally, soils were covered with perennial vegetation or annual species that re-seeded themselves every year. After European settlers arrived in North America, this vegetation was largely replaced with annual crops, such as corn, wheat, cotton and, later, soybeans, on the most productive soils. Traditionally, fields of these crops have been plowed after harvest in the fall or during the following spring and then have been seeded again with an annual crop. This process leaves the soil bare for a month to several months each year. During this time the soil is subject to severe erosion by water and wind. The traditional agricultural environment is greatly different from the natural environment, and, as such, it is inviting problems. According to the first general principle of soil care stated at the beginning of this chapter, the soil should have a surface cover at all times.

### *Biodiversity*

Soils in their natural states support a great diversity of plant, animal and microorganism species. For example, if one plant species is devastated by disease or insects, the population of the host plant will be reduced and the population of the disease or insects will be increased for a time. The overall effect on the entire plant community, however, will be small because the susceptible plant species constitutes only a small portion of the entire community.

In most farming operations, often only one species is planted (monoculture), and that same crop might be grown year after year. This provides an ideal environment for weeds, insects and diseases that prey on the crop to build up large populations. The first general principle of soil care tells us to strive for a mixture of plant species. We can achieve some degree of biodiversity by growing a mixture of species, such as in forage crops, and by following a crop rotation in which different crops are grown in successive years. Crop rotations limit the build-up of a plant pest by removing its host plant from a field for several years. They also reduce the need to chemically control the pests.

## *Purification of Waste Products*

Soils purify the *natural* waste products added to them. Over thousands of years of soil formation, organisms evolved to break down wastes (such as those from animals) into simpler products like soil organic matter, carbon dioxide and water. These products can cycle back into the atmosphere or into ground and surface water, or they can be stored in the soil.

Although soil organisms can break down and purify natural materials, they might not be able to process chemicals that were not part of the environment during soil formation, such as synthetic chemicals. Some herbicides and insecticides, for example, persist a long time in soils before they are broken down. Other chemicals, such as heavy metals like mercury, lead and cadmium, are held tightly by soils and, once added, remain in them practically forever. In general, any chemical not natural to the soil should be added with care.

## *Influence of Soils on Climate*

We learned earlier how climate influences soils. Soils also influence the climate. They moderate daily and yearly climatic fluctuations and those related to weather cycles. During a rain storm, some rain is absorbed by the soil and usually some runs off the surface. The absorbed rain is used gradually by plants during times of no rain. The runoff can cause soil erosion and flooding. Therefore, absorption of water by the soil greatly lessens the devastating effects of a rainstorm.

Because heat moves through soil slowly, soil heats up slower than air. This effect is shown in both daily and seasonal temperature cycles. During the day soils absorb heat and at night they radiate it, thus reducing extreme daily temperature fluctuations. Dark-colored soils absorb more heat during the day and radiate more at night than light-colored soils. For this reason, farmers select dark-colored soils for growing crops that need warm evening temperatures. Evaporation and transpiration processes absorb heat and tend to cool their environment. In cities and towns landscaping with trees, grass and bushes, as compared with inorganic materials such as concrete and rock mulches, moderates soil and environmental temperatures.

Retardation of heat flow in soils is especially evident between summer and winter. In the summer, subsoil temperatures are generally cooler than air temperatures, and during the winter subsoils are warmer. The time of the annual temperature maximum at a depth of 2 meters is several weeks later than at the surface, and the minimum has a similar lag. That is why in northern latitudes water pipes in the soil sometimes freeze at about the time the air temperature begins to rise, in the late winter months. These soil-temperature relations also affect how plants grow. In the spring, plant growth is often slower than one would anticipate from the air temperature because the soils are cooler, but in the fall, growth of plants such as pasture grasses continues well into the cold season.

Principles of heat flow in soils can also be used for designing home sites. Some buildings are earth-sheltered, built largely underground, to reduce the temperature fluctuations around them and to conserve fuel for heating and cooling. Geothermal heating systems utilize the seasonal lag in temperatures in deep soil layers to reduce energy use. Air conditioning systems usually remove heat from a building by exchanging it with the outside air. On hot days, the outside air is not very effective in removing heat, because it has already absorbed a great deal of heat. The soil at a depth of a few meters, however, is much cooler and can absorb more heat. In the summer, geothermal heating-cooling systems transfer heat from the building to this cool soil by circulating liquids through pipes buried in the soil. This same system is used to remove heat from the soil in the winter, when the soil is warmer than the outside air, and to transfer it to the building.

Soils indirectly influence global climate change because they play a key role in the carbon cycle. Increased carbon dioxide ($CO_2$), as well as other gases, in the atmosphere has been blamed for global warming. Most of this extra $CO_2$ has come from burning fossil fuel, such as coal and oil, but some has come from the soil when organic matter is oxidized to $CO_2$. Live plants reverse this process removing $CO_2$ from the atmosphere and fixing it in plant tissue in the process of photosynthesis. Some of this plant material is returned to the soil. To help lower the $CO_2$ content of the atmosphere, it is necessary to maintain vigorous growth of vegetation and to promote high levels of organic matter in the soil. Both processes are encouraged in a healthy soil. Furthermore, as mentioned in chapter 11, good soil cover reduces erosion of the surface soil layers which contain the most organic matter.

## Wetlands

Soils that are saturated with water to the surface for several months a year are called hydric soils, and the entire ecosystem associated with these soils is called a wetland. Wetlands replenish ground water, and they serve as habitats for wildlife such as ducks, geese and other animals. Wetlands were extensive in the United States when European settlers arrived, but since then most of them have been drained by underground tiles or tubes and ditches for agricultural production. These drained wetlands comprise some of our most productive soils because they hold much water and supply it for plant growth during the dry summer season. Without agricultural production from them our food supply would be limited.

Because wetlands are such a vital part of our entire ecosystem, we must preserve those that remain in their natural state. In fact, federal legislation prohibits draining most of our remaining wetlands. Wetlands are identified by the length of time they are saturated with water and the kinds of vegetation they support. In most hydric soils, the water table level fluctuates with weather patterns and with seasons of the year, so the soil might not be saturated year-round. Soil wetness leaves its mark on soil morphological features, as explained in chapters 7 and 8, so it is possible to see the results of wetness and chemical reduction even when the soil is not saturated.

## Soil Degradation

A healthy soil is one that is well adjusted to its environment. Deviations from the healthy condition result in soil degradation. Any form of soil degradation also degrades the environment at that site. Furthermore, some kinds of soil degradation affect the environment in other places, and they may affect the global environment. Soil degradation processes, and how they might be countered, are discussed in this section. One kind of soil degradation, soil erosion, is so pervasive that it is covered separately in chapter 11.

### Soil Structural Decline

*Soil structure* refers to the geometric manner in which soil particles are grouped together to form aggregates or peds. Conversely, *pore space* refers to the space not occupied by solid soil particles. Many soil processes depend on the availability of water and oxygen in the soil. Plant roots, microorganisms, insects, and soil animals all need air and water which move through the soil pores. In soil with good structure, particles are bound together in stable aggregates, and this results in a system of pores through which water, air, and roots can readily move. Most soils had good structure before they were first tilled, but many farm operations destroy these aggregates.

The main manifestations of structure decline are compaction, sealing and crusting. Compaction results from the mechanical compression of soil particles and aggregates by the weight of farm equipment on the soil and by the force exerted by tillage implements. Surface horizons are subject to compaction from heavy implement traffic on the soil surface. Also, when soils are plowed, the zone immediately below the plow layer is compressed by the weight of the tractor in the furrow and by the force exerted by the moldboard plow. This compacted zone is called a *plow pan*. Soils are most subject to compaction when they are moist. Compaction is minimized by reducing the number of trips over the soil and the number of tillage operations, especially when the soil is moist.

Surface sealing takes place when raindrops hit bare soil. Individual particles are detached and they move downward and plug up the pores in the soil. This "washed-in" layer is typically a few millimeters thick. At the very surface, a compacted zone of less than one millimeter forms by raindrop impact. These layers comprise the seal. The seal is almost devoid of pores and causes greatly increased water runoff. When the seal dries it becomes very hard and is called a *crust*. Crusts prevent seedling plants from emerging through the soil surface. Sealing and crusting are also minimized by maintaining a surface cover, which reduces the impact of raindrops on the soil. Also, reduced tillage results in less pulverization of soil aggregates and less sealing and crusting.

### Leaching and Acidification

When minerals weather, soluble cations, such as the sodium ion ($Na^+$), the calcium ion ($Ca^{2+}$), the magnesium ion ($Mg^{2+}$), and the potassium ion ($K^+$), and soluble anions, such as chloride ($Cl^-$), carbonate ($CO_3^{2-}$), bicarbonate ($HCO_3^-$) and sulfate ($SO_4^{2-}$), are released first. Some of these ions are essential plant nutrients. In humid areas the anions are quickly leached from the soil, but cations are held by the negatively charged sites on clay and organic matter. Eventually, however, cations are replaced by the hydrogen ion ($H^+$) and leached out, which makes the soil more acid. Some of this acidity is converted to aluminum compounds (instead of $H^+$), which, in high concentration, are toxic to many plants. These leaching processes are a natural part of soil formation, but they are accelerated by acid rain -- precipitation that is made more acid by sulfuric and nitric acids from smokestacks and automobile exhaust. Nitrogen fertilizer also creates some soil acidity. In farming, soil acidity is counteracted by adding lime ($CaCO_3$), and the loss of other nutrient elements is compensated by adding fertilizer. Some of these soil amendments do not move readily through the soil, so low fertility and high aluminum contents in subsoil horizons are especially hard to correct.

## *Salinization*

Salinization is the accumulation of excess salts in a soil. It is the opposite of leaching. In arid and semi-arid areas, there is not enough rainfall to leach out the salts (soluble cations and anions) and they accumulate in the soils. In some arid areas the soils are naturally high in salts, and in other areas salinity problems have increased because of human activity. If the salt content is not too high, these soils support certain salt-tolerant species. In large parts of the world, especially in semi-arid areas, the ground water has a high salt content, but it is below the rooting zone, so the salt has little effect on plant growth. Problems arise, however, if the ground water level rises. The water table can rise when trees are cut and replaced with annual crops that transpire much less water. In Australia, for example, native eucalyptus trees transpire large amounts of water because they keep their leaves all year. Many of these trees have been replaced by annual crops, such as wheat, that use much less water. The water not used has moved downward and has caused the saline ground water level to rise into the rooting zone, where it has retarded plant growth. In some of these areas trees are being replanted to lower the water table so crops can be grown in nearby fields.

As stated above, soils prone to salinization are found in arid and semi-arid areas, which are likely to be irrigated. Irrigation presents special problems. If the water table is deep but salty and the soils are permeable, special care should be taken to avoid over-irrigation, which could raise the water level into the rooting zone. If the soils conduct water slowly or have a shallow water table, an underground drainage system should be installed so the soil can be flushed out occasionally. In addition to containing their normal salt content, these soils are usually fertilized, which adds even more salt and increases the need for drainage.

Drainage of saline soils may cause problems for farms downstream along a stream used as a source of irrigation water. The irrigation water used at the first farm picks up salts from the saline soil and from fertilizers as it leaches through the profile. The excess irrigation water drains back into the river and is used by the second farm. The farther downstream the water travels, the saltier it becomes.

Soils can also become salty in humid coastal areas, where rainfall exceeds evapotranspiration and the surplus water leaches through the soil to the ground water and eventually moves to the ocean. In these areas, however, if much ground water is pumped from the wells, the direction of flow is reversed: salty ocean water replaces fresh ground water and eventually salt water is pumped from the wells. If this water is used to irrigate soils, they eventually become saline.

## *Desertification**

The term *desertification* is pictured by some to represent the physical process of the desert encroaching on more humid areas -- that is, the process of great seas of sand dunes covering grassy areas. Actually, desertification implies a complex interaction of humans and nature. Desertification is most severe in northern Africa. It is not happening so much at the very edge of the desert as around population centers that may be some distance from the desert. Many of these areas already have more people than the resources can support. In wetter years the soils are sufficiently productive to support the population. During a drought, however, people remain on the land in the hope that the rains might soon return. They continue to graze their goats, sheep, cattle and camels on the withered grass, and when it plays out, they feed the animals by trimming trees already weakened by the drought. The people also go on collecting firewood from the sparse trees and shrubs. Eventually the vegetation dies, and the unprotected soil blows away. When the rains do occur, severe water erosion occurs, and the whole area is less productive than it was during the previous, wetter period. This cycle continues, with more and more damage to the environment occurring with each drought. Daniel Hillel explains the process as follows:

The damage to vegetation and soil typically results from the combined effects of the prolonged drought and the excessive pressure by too many people and animals on a

---

*This section is based largely on the book by Daniel Hillel, *Out of the Earth* (Berkeley: University of California Press, 1991).

land too parched to support them. The same land that supported them and their forebears for so many generations is thereby in effect turned into a desert. The tragedy of the American Dust Bowl is thus being reenacted in Africa on a vastly greater scale of geography and human suffering. (p. 191)

## Soil Contamination

In its natural state, soil purifies water on its way to an aquifer or to a stream by absorbing and gradually breaking down potential pollutants. A soil can become contaminated, however, and add impurities to the water rather than removing them. These impurities may be excessive farm chemicals, residential or industrial waste products, or toxic materials buried in landfills.

In a proper fertilization scheme, farmers add plant nutrients to the soil at approximately the same rate the nutrients are removed from the field through harvested crops. If they add more nutrients, the soil might absorb some of them, but eventually it will pass them on to other parts of the environment. Mainly, fertilizer nutrients are added to the soil either as chemical fertilizers or as manure. The compositions of chemical fertilizers are well known and stated on tags, and fertilizer spreaders apply them accurately and uniformly. Manure, on the other hand, is applied mainly to get rid of it, little is known about its composition, and manure spreaders are rarely calibrated. Because of these differences, plant nutrients in manure, especially nitrogen and phosphorus, are often applied at very high rates -- much higher than can be used by crops. Manure from large cattle, swine and poultry operations might need to be transported long distances if it is to be applied at proper rates.

Nitrogen and phosphorus act quite differently in soils. Nitrogen often occurs in, or is converted to, the nitrate form ($NO_3^-$), which is highly soluble and moves with the soil water. From the soil, nitrate can move into ground water and into the water that flows from drain tiles into surface waters. Water too high in nitrate is not healthful to those who drink it; it can cause illness, and even death, to infants. Phosphorus, on the other hand, is held tightly by soil particles and causes damage mainly through soil erosion. If, for example, soil high in phosphorus is eroded and is carried to a lake, the phosphorus held by the soil will become available to the aquatic plants. Aquatic plants flourish in the presence of phosphorus and, through a process called eutrophication, begin to consume nearly all the oxygen in the lake.

Most herbicides, insecticides, and other chemicals added to the soil are broken down into simpler, nontoxic products by soil microorganisms. Some of these chemicals, however, are not readily degraded. Those loosely held by soil particles may get into water supplies and eventually into the food chain, where they may be ingested by animals and people. Some chemicals that were heavily used in earlier years, such as DDT and chlordane, have been shown to be dangerous to people and the environment and have been banned from further use. Other chemicals are thought to cause cancer and other diseases, so there is much concern about their use.

Before the hazard of some farm chemicals was known, there was a tendency for many farmers to treat whole fields with a pesticide -- just in case a certain weed, insect or disease might become a problem. In recent years there has been much emphasis on using fewer pesticides. Their use is being reduced by *site-specific* management practices, whereby a person trained to identify weeds, insects, diseases, soil compaction and other problems investigates a field and recommends which chemicals or alternative controls should be applied to certain areas of the field.

## Landfills

Another source of soil contamination is from commercial and residential waste products in landfills. Many older landfills are in soils that are subject to leaching, often on terraces and flood plains near rivers, and the leachates get into the ground water and surface water. Stricter regulations are being imposed on how landfills are designed, constructed, filled and maintained to reduce the potential for contamination of ground water. These landfills will be much more expensive to operate than previous ones have been. Thus, in addition to being environmentally wise, it will be more cost-effective to recycle more of the things we formerly threw away.

## Sewage Disposal

Some soils are used to receive and purify human wastes. In municipal disposal plants, sewage undergoes treatment, but some of it, the sludge, is not broken down into simpler materials and must be disposed of somewhere. One alternative is to burn it, but this requires fuel and contributes to air pollution, therefore, much of this sludge is applied to soils. Soil microorganisms can break down most sludge that originates from residential sewage. However, some industries dispose of potentially toxic substances in the sewer system, and they also become part of the sludge. Industrial sludge may contain heavy metals, such as cadmium, lead, zinc and copper, which potentially can get into plants grown on the soil in unsafe levels. Industries are now being much more careful about what they put into sewers than they were in previous years.

Beyond the limits of municipal sewer systems, people depend on individual sewage disposal systems. Most utilize a septic tank, which provides primary treatment. The effluent from the tank passes through drain lines and is absorbed by the soil, usually by the subsurface horizons. Aerobic soil microorganisms (those requiring air) break down the effluent into harmless materials. If the soil horizons in which the effluent lines are placed are well aerated and moderately permeable, the effluent remains in the soil long enough for the soil organisms to break it down. The system fails, however, in waterlogged, poorly aerated soils in which the aerobic microorganisms are not active; in very highly permeable soils through which the effluent passes so rapidly that it is not broken down; or in very slowly permeable soils that do not absorb effluent as fast as it is supplied from the septic tank. Therefore, it is important to investigate soils thoroughly before a waste treatment system is installed.

## Concerns of All People

Many of the problems of soil degradation must be solved by those who farm and manage natural resources, but public policy sets the guidelines for how farmers and managers use their soil resources. We all have an input in developing those policies. Some aspects of public attitude and policy affect all people and also affect our soil resources. Two of these concerns are addressed here.

## Population Pressures

The connection between the size of the population and the quality of our soil and water resources was clearly stated by Paul R. and Anne E. Ehrlich in 1990[*]:

> The size of the human population is now 5.3 billion, and still climbing. In the six seconds it takes you to read this sentence, eighteen more people will be added. Each hour there are 11,000 more mouths to feed; each year more than 95 million. Yet the world has hundreds of billions *fewer* tons of topsoil and hundreds of trillions *fewer* gallons of groundwater with which to grow food crops than it had in 1968. (p. 9)

The growth of the population is illustrated by the time it takes to double the population of the world (figure 12-1). The total human population at the time of Christ was around 200 million to 300 million people, and it increased to about 500 million by 1650. It then doubled to 1000 million (1 billion) around 1850, and doubled again to 2 billion by 1930, and once again to 4 billion by 1975. From another perspective, it took from the beginning of human-kind to the year 1650 to add the first half-billion people. Then it took 200 years to double the population to 1 billion, 80 years to double it to 2 billion, and 45 years to double it again to 4 billion. The population now stands at 5.6 billion, and according to current growth rates, it will reach 8 billion by the year 2020.

---

[*]Paul R. Ehrlich and Anne E. Ehrlich, *The Population Explosion* (New York: Simon and Schuster, 1990). Other parts of this section are also based on that book.

The sheer *numbers* of people is not as important as the *impact* these people have on the environment. The Ehrlichs devised an equation to illustrate this impact:

$$I = P \times A \times T$$

where I is the impact on the environment, P is the population, A is a measure of the average person's consumption of resources (which is also an index of affluence), and T is an index of how much environmental harm is caused by the technologies that provide the goods consumed. This equation illustrates some important points. First, it shows that a baby born in the United States has more of an impact on the environment than one born in Africa, for example, because the A factor is so much larger in the United States. It also explains why a little development in poor nations with big populations, like China, can have an enormous impact on the planet. A very large population (P) multiplied by a small increase in consumption (A) results in a large increase in the impact on the environment (I).

Are current populations too high? Human ecologists try to answer this question by referring to the *carrying capacity* of a region. The carrying capacity of a region is the population the area can sustain without swift exhaustion of its nonrenewable resources (such as soils, fossil fuels, metals and water in deep aquifers) and damage to its renewable resources (such as ground water, rivers and forests). When the demands of people surpass the environment's carrying capacity, there are too many people. Human ecologists believe that the carrying capacity has been exceeded in many parts of the world.

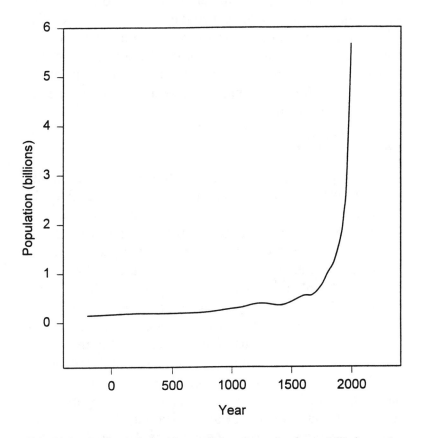

Fig. 12-1.   Estimated world population from the time of Christ to the present. (Plotted from data in figure 6.2 of *The Atlas of World Population History* by C. McEvedy and R. Jones, 1979, New York: Facts on File, and from data from the Population Reference Bureau, Washington, DC.)

The rate of population growth differs greatly among countries. It is generally much higher in the less developed regions, such as Africa, Latin America and southern Asia, and lower in the more developed regions, especially Europe and North America (with the rate in the United States being higher than in most European countries). However, we live in a global environment, where population growth in one part of the world affects the quality of the soil and water, and other phases of the environment, in all parts of the world. Therefore, population pressures should be of concern to all people.

## *Land-Use Planning*

Another concern of all people should be how we use our land resources, and the *amount* of land available for farmland, forest land and natural areas. Once land is removed from use by houses, roads and industrial development, it is rarely allowed to return to a natural or productive state. Today, the United States is losing 3.5 acres of farmland every minute to encroaching urban sprawl. People in Europe, and other places, have been managing their land resources for a much longer time than have Americans. They have learned that if future generations are to inherit healthy natural areas and productive farmland, these areas must be intentionally preserved. They have developed a system of compact "development" of residential and commercial areas, leaving the countryside open. In contrast, development in the United States has resulted in cities with large, poorly used urban areas, sprawling suburban areas, strip development along highways, and large areas in the countryside devoted to single residences. Interest in preserving farmland and natural areas is increasing in the United States. We can learn much from our international neighbors about land-use planning to foster this interest.

## Summary

In earlier chapters, it was shown that soil and ecosystem properties are a function of the factors under which the soils formed: parent material, topography, climate, organisms, and time. This chapter emphasized the relation of soils to their *present* environments. Two general principles were stated:

1. Soils existed for a very long time under certain natural conditions; if we plan to use them for a long time we should emulate those conditions as closely as possible.

2. During the formation of soils, properties changed very slowly. Many properties still change slowly in response to our use of soils, so it is possible that we are slowly damaging soils and not noticing it. Therefore, we must be sensitive to signs of degradation, or else the soils will become so damaged that it will be difficult or impossible to reverse the harm.

This chapter also reviewed some factors of the present environment that are important to how we use our soils. Next, we discussed the attributes of a healthy soil and how deviations from this standard can lead to various kinds of soil degradation. Finally, it was pointed out how common concerns of all citizens -- population pressures and land-use decisions -- affect our ecosystem, of which soils are a key component.

**absorption** the movement of water and ions into the plant root.

**adsorption** the process by which ions or molecules are held on the surfaces of soil particles.

**aerated soil** a soil in which the air is similar in composition to the air above the soil.

**aerobes** organisms that grow only in the presence of oxygen.

**aggregate** many soil particles held in a single mass or cluster, such as a clod, crumb or block.

**anaerobes** organisms that grow in the absence of oxygen.

**available water capacity** the amount of water held in a soil that is obtainable by plants; the difference between field capacity and permanent wilting point.

**cementation** the process in which soil particles are held together by certain chemical materials.

**clay** (a) Mineral soil particles less than 0.002 millimeter in diameter. (b) A soil containing more than 40% clay.

**clay skins** coatings of clay on the surfaces of soil peds, mineral grains or in pores.

**complexation** a kind of chemical reaction in which an organic compound bonds to an atom such as iron or aluminum.

**delineation** an individual area enclosed by a line on a soil map.

**denitrification** the process of reduction of nitrate ($NO_3^-$) to gases such as nitric oxide (NO), nitrous oxide ($N_2O$), or nitrogen gas ($N_2$); takes place in water-logged soils.

**detachment** the part of the erosion process that removes individual soil particles from an aggregate or ped mainly by the impact of rain drops. These particles are then moved by the transport process.

**diagnostic horizons** kinds of soil horizons defined to aid in classifying a soil in Soil Taxonomy.

**erosion** the wearing away of the land surface by wind, water, gravity or a combination of these forces.

**evapotranspiration** the sum of evaporation and transpiration.

**exchangeable cations**  positively charged ions held by negative charges on the surface of a soil colloid (small particle) such as clay or organic matter.

**field capacity**  the water content of a soil after it has drained a few days following complete saturation.

**foliar analysis**  chemical analysis of plant parts in the laboratory to assess the nutritional status of the plant.

**fragipan**  a subsoil layer that restricts water movement and root growth.

**gilgai relief**  surface topography characterized by small knolls and small depressions.

**ground water**  subsurface water in the zone of saturation.

**gully erosion**  a process in which water flows in channels and removes soil to depths more than 20 inches.

**horizon**  a layer of soil, approximately parallel to the surface, that has distinct characteristics produced by soil forming processes.

**horizon designation**  a set of letters and numbers used to represent the major properties of a soil horizon.

**humus**  the dark-colored soil organic matter that has fairly definite chemical and physical properties and is not subject to as rapid decomposition as plant residues.

**hydrologic cycle**  the fate of water from the time it falls as precipitation until it is returned to the atmosphere through evaporation or transpiration.

**hydrology**  the science dealing with the water cycle in nature.

**immobilization**  the process in which inorganic forms of nitrogen, such as nitrate ($NO_3^-$) and ammonium ($NH_4^+$), are incorporated into the tissue of an organism.

**inclusion**  a kind of soil in a soil map delineation that is different from the main or named kind of soil in that delineation.

**infiltration**  the movement of water into the soil through the soil surface.

**landslide**  rapid downslope movement of a mass of soil, mainly due to gravity.

**leaching**  the removal of soluble material from soils by percolating waters.

**liming**  the process of applying lime (mainly calcium and magnesium carbonates) to the soil.

**loam**  a soil containing similar amounts of silt and sand and a smaller amount of clay.

**map unit**  a collection of the same kind of soil delineations.

**map unit symbol**  the alphanumeric symbol used to represent a certain kind of soil map unit.

**microbes**  organisms not visible to the naked eye; microorganisms.

**mineralization**  the conversion of an element such as nitrogen from an organic form to an inorganic form such as nitrate ($NO_3^-$) and ammonium ($NH_4^+$).

**mottled pattern**  spots of one soil color interspersed in another color. Gray mottles usually indicate waterlogging in the soil.

**nitrification**  the process of oxidation of ammonia ($NH_3$) or ammonium ($NH_4^+$) to nitrate ($NO_3^-$); takes place in well-aerated soils.

**nitrogen fixation**  the conversion of nitrogen gas ($N_2$) in the atmosphere to nitrogen-containing organic compounds.

**organic matter**  soil material derived from living material, composed of carbon-containing compounds.

**oxidation**  the process in which an atom loses an electron, often by combining with oxygen.

**parent material**  the geologic material from which soils form.

**ped**  an association of soil particles combined into a unit or aggregate such as a granule, block, prism, or plate.

**pedon**  a three-dimensional body of soil large enough to study its horizons.

**permanent wilting point**  the water content of a soil so dry that plants growing in it wilt and do not recover.

**photosynthesis**  the process in which the energy of sunlight is used by green plants to create organic compounds and oxygen from carbon dioxide and water.

**plant residues**  decayed plant parts that remain in the soil.

**plant tissue testing** chemical analysis of plant parts in the field to assess the nutritional status of the plant.

**pore space** that part of the total volume of the soil not occupied by solid particles.

**reaction** the degree of acidity or alkalinity of soil, frequently expressed in pH units.

**reduction** the process in which an atom gains an electron, usually in the absence of oxygen.

**respiration** the process in which plants utilize ("burn") organic compounds and oxygen and give off carbon dioxide and water.

**rill erosion** a process in which water flows in numerous small channels and removes soil to depths of less than 20 inches.

**salt** a chemical compound characterized by a basic part and an acid part. In table salt, NaCl, Na represents the basic part (e.g., NaOH) and Cl represents the acid part (HCl).

**sand** (a) Mineral soil particles between 0.05 and 2.0 millimeters in diameter. (b) A soil made up predominantly of sand.

**saturated** the condition when all (or practically all) of the soil pores are filled with water.

**sheet erosion** the removal of a uniformly thin layer from the soil surface.

**shrink-swell** the process in which parts of a soil (peds) become smaller by drying or larger by wetting.

**silt** (a) Mineral soil particles between 0.002 and 0.05 millimeter in diameter. (b) A soil made up of 80% or more silt.

**soil** (a) The unconsolidated cover of the earth, made up of mineral and organic components, water and air and capable of supporting plant growth. (b) A natural body, occurring in various layers, composed of unconsolidated rock fragments and organic matter.

**soil consistence** a field estimate of soil strength; for moist soil, consistence is described (from weak to strong) as loose, friable, and firm.

**soil formation factors** the factors that determine the kind of soil that forms; they are climate, organisms, parent material, relief (topography), and time.

**soil forming process** a mechanism or course of action by which a soil forms.

**soil interpretation** the process of using soil information to predict the suitability or limitation of soils for various uses.

**soil matric potential** a measure of how tightly a soil holds water, or the energy required to remove water from the soil.

**soil moisture regime** (a) The moisture condition of a soil through the year, including the depth to the water table. (b) The specific soil moisture classes defined in Soil Taxonomy.

**soil morphology** the physical constitution of the soil, including color, texture, structure, consistence, nature of ped surfaces and other properties.

**soil order** the highest category in Soil Taxonomy.

**soil profile** a vertical section of the soil through all its horizons.

**soil series** the lowest category in Soil Taxonomy.

**soil structure** the combination of primary particles into aggregates or peds. The major shapes of peds are: granular--roughly spherical shape; blocky--block shaped, approximately equal in horizontal and vertical dimensions; platy--plate shaped, horizontal dimension grater than vertical dimension; prismatic--vertical dimension greater than horizontal dimension.

**soil survey** a document containing a systematic examination, description, classification and mapping of soils in an area.

**Soil Taxonomy** the current system of soil classification in the United States. *Soil Taxonomy* is the name of the book that describes the system.

**soil temperature regime** (a) The temperature of a soil through the year. (b) The specific soil temperature classes defined in Soil Taxonomy.

**surface area** the quotient of the total surface of the soil divided by its dry weight (usually expressed in square meters per gram).

**surface runoff** water from precipitation discharged from an area by flowing over the soil surface.

**texture** the relative proportions of sand, silt and clay in a soil.

**tilth** the physical condition of a soil in relation to its plant growth.

**transpiration**  the process by which water is evaporated through plant leaves.

**transport**  the part of the erosion process that involves movement of soil particles away from the point of detachment.

**water requirement**  the mass of water required to produce one unit of dry plant material.

# INDEX